INTO THE HEART OF OUR WORLD

INTO
THE
HEART
OF OUR
WORLD

A Journey to the Center of the Earth:
A Remarkable Voyage of Scientific Discovery

DAVID WHITEHOUSE

PEGASUS BOOKS
NEW YORK LONDON

INTO THE HEART OF OUR WORLD

Pegasus Books Ltd.
80 Broad Street, 5th Floor
New York, NY 10004

ISBN: 978-1-60598-959-4

10 9 8 7 6 5 4 3 2 1

Printed in the United States of America
Distributed by W. W. Norton & Company, Inc.

To Bernard Carey and Derek William Whitehouse.
In Memoriam.

Contents

List of Illustrations

'I am scarcely able to believe in the reality of my adventures. They were so truly wonderful that even now I am bewildered when I think of them . . .

'This is the final conclusion of a narrative which will be probably disbelieved even by people who are astonished at nothing.'

Jules Verne, *Journey to the Centre of the Earth*

Introduction

Throughout my life I have always looked outward contemplating the view from planet Earth. My life as an astronomer took my gaze to nearby stars and distant galaxies as I used telescopes and satellites to observe the cosmos. Rarely did I think about the ground beneath my feet. I knew the broad outline of how the Earth had been born, certainly, but for me it was a platform and not a place for study.

Having written books about the Moon, the Sun and about space flight, as well as a biography of Galileo, I was looking for a new topic and there was no shortage of ideas. I knew that many of them would not interest me for the length of time needed to write a book so I searched wider and started reading, after many years of neglect, classic works of science fiction, and I was led inevitably to the feet of H. G. Wells and Jules Verne.

Both were visionaries and storytellers but Verne had a streak of adventure that excited me and when I discovered that perhaps his most famous work, *Journey to the Centre of the Earth*, was soon to celebrate its 150th anniversary I began to look at the topic of the Earth somewhat more closely and realised what I had been missing. I had studied other worlds in space, strange worlds, surprising worlds, but none as surprising as the worlds that reside in our world. If you want strangeness and surprises, look below.

Many of my friends have stories of how, as children, they discovered the night sky and with a small telescope and star map began exploring it. Now I was talking to people who as children collected rocks and wielded a geological hammer as I had a telescope. I had the stars in my mind and on my charts, but they had their rocks, fossils and crystals in their hands.

So now, every time I look up at the night sky a part of my mind is left behind on the planet where I live and in places I can never go. I can look at images of the surfaces of Venus and Mars, even the Sun, but I will never have such images of the inside of our planet. There are places down there that are astounding and significant. Astronomers often say we are made of stardust and are the children of the stars but the Earth is no less our parent. There are out in space stars that no doubt have planets circling them and some of those planets will be like our Earth with intelligent creatures living on it. Some of those stars will have been scattered from their parent galaxies, condemning their passengers to an almost eternal isolation. With no prospect of leaving I wonder if they appreciate their world, and their links to its interior, better than most of us humans do.

The voyage to the heart of our planet has science as the navigator. But it is not a voyage we undertake alone. From the mystic dreamers of past millennia to the first visionaries and the practical scientists it is not just a story of rocks, minerals and atoms. It is about people, conflicts and tragedy, discoveries and despair, for every earthquake that brings ruin and death also shows the route we must take but can never travel.

But aren't the best journeys the ones we can never make?

On my journey I thank the following for their advice: Don Anderson, Jonathan Aurnou, David Boteler, William Butcher, Judith Coggon, Edward Garnero, Dan Frost, Cornelius Gillen, Steven Hauck II, Dan Lathrop, Jeffrey Love, Karen Lythgoe, Maurizio Mattesini, Jonathan O'Neil, Wayne Richardson, Lisa Rossbacher, Robert Stern, Dmitry Storchak, Hrvoje Tkalcic and John Valley. I also want to thank Penny Armstrong, Nick and Sarah Booth and Pippa Cox.

I also profoundly thank my family, Jill, Christopher, Emily and Lucy and William Edwards. My agent, Laura Susijn, believed in this book when it was just an idea and kept me at it even when I was flagging. I also thank Alan Samson of Orion Books for championing and editing the book.

1

The Archive of the Earth

On 8 February 1828, Jules Gabriel Verne was born the first of five children to Pierre and Sophie Verne. His father was a Parisian lawyer and his mother was of Scottish and Breton descent. Despite there being no literary heritage in his family, Jules wrote a great deal in his youth but it wasn't until he reached the age of thirty-five that he brought out his first novel, *Five Weeks in a Balloon*. It was a great success. He followed it up with *Paris in the 20th Century*, which, to his surprise, was rejected by his publisher with harsh criticism. It would not be published until 1994. So, despite his initial success, he felt the pressure to produce another work that was acceptable to his publisher.

Verne was living in Paris, having married Honorine de Viane Morel in 1857. Despite his interest in science he did not mix with scientists; throughout his long career of writing science fiction stories he never did. He had a magpie mind, aware of the political and scientific debates of the age – the age of the Earth, the nature of the stars and the evolution of mankind. He was intrigued by a book written by Scottish geologist Charles Lyell called *Principles of*

Geology, written in 1830. It was a landmark in geological science because it said that the processes that shaped the world in the past are the same as those that operate in the present. For this to happen long spans of time were required. The book was a major influence on the young Charles Darwin who was given a copy by Robert FitzRoy, the captain of HMS *Beagle*. Darwin later said that he had looked at rock formations 'through Lyell's eyes' and saw them in a new perspective. Although the second of Lyell's three-volume series of books rejected evolution, by the time Darwin read it he was already well on the way to formulating his own views on the subject.

Lyell's book disrupted mankind's relationship with time that had been built up over the centuries. According to the Bible the Earth was 6,000 years old and God had created mankind shortly after he had made the Earth. Lyell showed that the Earth was at least millions of years old and in rewriting the past Lyell enabled writers like Verne to reshape the future. The almost unimaginable spans of past time were matched by the times to come that would enable mankind to scale new heights of technology, achievement and discovery.

In the frontispiece of the first volume of *Principles of Geology*, there was a magnificent cross-section drawing of a volcano, which must have impressed Verne. He also read books that flowed from Lyell's insight, such as those written by French scientist and writer Louis Figuier (1819–94), who had published *The World before the Deluge* in 1863. Darwin's *On the Origin of Species* had come out only a few years earlier. It set Verne thinking.

Borrowing ideas from Figuier rather more obviously than he should, or would be allowed today, he produced *Journey to the Centre of the Earth*, which has become a classic. He wanted to show that the Earth is not 6,000 years old and that the devil does not live underground. The Christian chronologies clashed with the geological timescales of Lyell and Verne, who preached their sermon with stones and not scripture, yet Verne, who described himself at the time as an 'intelligent, but orthodox Roman Catholic', does not allow his protagonists to talk of religious matters. In his work the stones do the talking.

Verne's works appealed on more than one level. They were adventure stories, straightforwardly told, that captured youthful imagination in a way that, say, H. G. Wells did not. But they were something more. Nature is being conquered by science, either through a journey deep underground, or in a voyage to the Moon, or 20,000 leagues under the sea. But while Verne had the scientific approach he did not have the science of the Earth. He knew nothing of atomic structure or radioactivity or the analysis of earthquakes when he delved into the archive of the Earth.

One of the three men who embarked on the journey to the centre of the Earth said that, 'For a scientist an unexplained phenomenon is a torture of the mind.' Ever since Jules Verne's day scientists have been discovering new things about the Earth, and their minds have been constantly tortured. I expect they always will be.

2

'Descend, bold traveller'

'Descend, bold traveller, into the crater of the jökull of Snaefell which the shadow of Scartaris touches before the Kalends of July, and you will attain the centre of the Earth. I did it.'

Arne Saknussemm

If you could dig a tunnel right through the Earth you could theoretically reach the centre of our planet in just twenty-one minutes. You can't, of course: such a tunnel would have to tolerate temperatures as high as the surface of the Sun and pressures three and a half million times greater than those at the Earth's surface. No material yet developed could withstand those conditions. But just suppose you *could* build it.

As your capsule was released at the top of the tunnel it would go into free fall pulled by the gravity of the Earth beneath it. In less than a minute you would pass through the Earth's crust – its outermost hard shell made of lighter rocks that is only 35 km thick and a mere 1 per cent of the Earth's mass. Racing shock waves from earthquakes, you reach the top of the largest part of the Earth, its mantle, comprising half of its volume and 68 per cent of its mass. You would pass through the denser rocks of the upper mantle where they start to flow like plastic and in a few minutes you reach a 660-km depth where an important transition occurs. Travelling ever down, you reach the base of the lower mantle in about eight minutes; this is one of the planet's most mysterious regions where strange structures reside and the wreckage of ocean floors descends, only to be recycled back to the surface by plumes of hotter rock that take many hundreds of millions of years to get back to the surface, if they ever do.

As you near the base of the mantle you must prepare yourself for the biggest shock of your journey. Nowhere in or on our planet is there such a dramatic change of scenery as the one you are about to experience. Suddenly, at a depth of 2,890 km, you burst through the rocky part of the Earth into a sea of liquid metal. This is the outer core, the size of the planet Mars. It occupies about 10 per cent of the Earth's volume but 27 per cent of its mass. For over 2,000 km you are a submariner in a sunless sea of swirling currents, slow-motion storms and cyclones of liquid metal riven with magnetic and electrical fields. Then another shock: after eight more minutes and 5,100 km down you plunge into a super-dense ball of solid iron and nickel with an undulating surface that looks like outstretched giant iron trees. This is the crystal core, only about half a per cent of our planet's volume, a little smaller than the Moon, but nearly 2 per cent of our planet's mass and perhaps the deepest mystery we will encounter, which you ponder as you pass giant iron crystals perhaps 100 km long. At the Earth's core the capsule would be travelling at almost 29,000 km per hour and at the point you reached the centre of the Earth you would be weightless. The trip from the core to the surface would be the reverse of the journey down as your weight returns and you hurtle towards what seems to be an infinite rock wall as you enter the lower mantle once again. Perhaps you would think of another way that scientists divide the layers of the Earth. You will pass through the asthenosphere, the relatively weak region at the top of the mantle where rocks can flow like plastic. Then, finally, the lithosphere, which is the rigid outer shell composed of the upper mantle and the crust which is broken into what are termed tectonic plates. You would decelerate until you reached the surface when you would stop. If one were being a little more realistic one would have to take into account air friction and pump out all the air from the tunnel.

Not that such travel directly through the centre of the Earth would connect many land masses. It is a little-realised fact that for most of the land on Earth the point on the opposite side is ocean. Indeed, one could classify the surface of our world as having an ocean hemisphere – the Pacific – and a land hemisphere. We shall see that the differing sides of planet Earth reflect what is going on at its surface and within it at the deepest level. For the USA it's the South Indian Ocean, for Africa the Pacific, although you could go from Spain to New Zealand and Chile to China. One could get over this problem by not having the tunnel pass through the very centre of the Earth. Curiously, this does not change the journey time of forty-two minutes to the other side. Dreams of travelling to the heart of the Earth are a phantasm. We can only scratch the Earth's surface and the deeper we go the stranger things become as we unpeel the planets within our planet.

At the start of my journey to the centre of the Earth I descended into a working deep mine. I am over a thousand metres underground at the bottom of Boulby potash mine in Cleveland in the north-east of England, one of the deepest mines in Europe. I am in a huge, dimly lit tunnel that stretches off into the distance along which power cables, air pipes and lights are strung. There are over 600 miles of tunnels down here, enough to reach London and back with a hundred miles to spare.

Rock salt crystals sparkle in the walls, which are hot to the touch. The obligatory luminous overalls and survival gear make me even hotter. Through my steel-toed safety boots I can feel the heat from the ground and my light overalls are already damp. Keep hydrated, said the rules, never go off on your own and never go to sleep.

Every so often a flatbed Ford passes, its headlights full on, and with metal cages holding workers, ferrying them to and from the face of the mine many kilometres out under the North Sea. Mostly, though, this is a silent world of occasional light and deep shadows. Once I turned a corner and saw a group of workers sitting at a wooden picnic table having lunch. The only illumination was from their helmet lamps. It looked as though they were suspended in space, in a starless, infinite blackness.

To reach the head of the shaft I had to go through three large steel doors, and an airlock that hissed. The descent in the mine cage was swift and smooth, the sides of the shaft racing past many times faster than in a conventional lift. As we descended back through geological time I made a mental note of the strata we were falling through. First there was a shallow layer of boulder clay deposited during the last Ice Age only 15,000 years ago. Then we quickly passed through a layer of ironstone only two million years old. Lias shales came next, laid down in a warm sea during the Jurassic era about 195 million years ago. Then we passed through the older Triassic-era Keuper Marl – an old name for mudstone and siltstone – and Butler sandstone followed by later Permian mudstone. Briefly we passed through the thin layers laid down during what was known as the great dying before finally reaching the Permian evaporates of 260 million years ago.

When we emerged over 1,000 m. down we entered a brilliantly lit hall; it could have been a factory on the surface. It was noisy because of the ventilation ducts and the conveyor belts that take the potash to the surface.

My guide suggested I turn my helmet light off. It was quite a shock. Few people ever experience true darkness, but this was it. To say it was black is to understate it. For a moment I felt as if nothing else existed except this dark, hot, strange world that I could not reach out to touch. Later I pressed my face against the side of the tunnel and listened. My guide must have wondered what I was doing. Earthquakes from half a world away were passing through these rocks every hour. Some of the shock waves had passed through the mysterious core of the Earth itself – my eventual destination.

In a way I am recreating the first part of an imaginary journey written a century and a half earlier by Jules Verne.

In Verne's novel *Journey to the Centre of the Earth*, Professor Lidenbrock and three companions travel to Iceland to find an empty volcanic shaft. Earlier Lidenbrock had discovered an ancient runic manuscript with an extra section inserted in code. When he deciphered it he was able to read an account of a journey into the Earth made by Arne Saknussemm centuries earlier.

On the volcano Lidenbrock found the passage. '"Here it is," gasped the Professor in an agony of joy, "here it is I – we have found it. Forward my friends, into the Interior of the Earth."' One of his companions, his nephew Axel, said that when they were quite ready to begin their descent, their watches read thirteen minutes past one.

It is amazing how swiftly you adapt, how quickly the world of air, space and sunshine becomes a memory as you accept your new surroundings. There was something almost comforting about being so deep underground. Somehow it felt simpler, more elemental, as if there was just me and the planet stripped of all its diversions. But I knew that was not true. The first lesson I have learnt is that the descent is not one into isolation. Each rock and mineral, each fracture and fault, every rumble and movement, even the heat, has its story. Like the pages of a book, with each page having taken a thousand years to write, the secrets of the strata can be read. The recent past, the geological past, and even those times before the Earth was born, have all left their imprint here, written in code. Everything is connected; all layers, regions and depths affect each other. You escape nothing by going underground. As we shall see, we all have a deep connection with the underworld. We are as much children of the core as we are the offspring of air and water.

If I were on this spot 260 million years ago I would be standing at the Earth's equator on the shoreline of the great shallow Zechstein Sea, watching it die. This sea, only tens of metres deep but a thousand kilometres wide, is hot under the baking sun. I can see water vapour rising from it, making the view out to sea misty, each departing molecule making what remains more salty, brackish and inhospitable to life. Here there are neither birds in the sky – for they are of future ages – nor animals nearby. The fish, reptiles and giant cockroach-like creatures that dominate life on this world are far away, living nearer the coast.

The vast swampy forests of the Carboniferous era are gone and the dinosaurs have yet to come. A fierce Ice Age and changing sea levels have recently altered the face of the planet and all the land is now massed into one great supercontinent called Pangaea, surrounded by a single global ocean called Panthalassa. Britain lies deep within the northern part of Pangaea in a land mass called Laurasia on the very edge of the salty Zechstein. Around it lies one of the world's great deserts, what's left over from the erosion of widespread barren uplands of previous geological epochs.

Things are changing on and in the Earth. Things are always changing on and in the Earth. The breakup of Pangaea is already under way, and its various parts have started their drift across the globe. Sooner than that, in ten million years or so, will come the great dying – the Earth's most severe extinction event when 95 per cent of all species are wiped out. No one knows the cause of it and some speculate that the release into the atmosphere of gases during the evaporation of the Zechstein Sea and others like it may have played a part. As the intra-continental sea shrinks it leaves behind vast flats of salts and minerals, particularly potash. The water is long gone. Its evaporative remains are buried near me, showing that in a way it is still possible to stand on the shore of this ancient ocean.

As I have said, we are not isolated here. The distant reaches of space and time can still touch us. Down one of the tunnels astronomers have set up experiments to look for the elusive dark matter that makes up most of the universe but which hasn't been identified. The theory is that dark matter is all around us, wandering through our planet from time to time, and can be detected by sensors placed under a kilometre of rock where the interference from other sources is minimal. When you understand the Earth it is not surprising that, en route to its centre, you can still feel the far reaches of time and space.

Potash from the Zechstein Sea is the treasure mined at Boulby. It is used as a fertiliser and Boulby produces half of the UK's needs. If you want something more precious you have to go deeper, about as deep as humans can go underground, and to the former shores of an ocean far, far older than the Zechstein Sea.

We dig into the Earth for many reasons. It's part of our nature to explore and it reveals itself as much in the child digging on a sandy beach as it does in the deepest hole ever drilled. The holes we have excavated into the Earth seem puny things. Take the Bingham Canyon copper mine located in Salt Lake County, Utah. It is currently the largest open-pit mine in the world at 4 km wide and almost 1 km deep. It has been in production since 1906. But keeping the hole intact isn't easy. In 2013 it suffered a landslide – the largest non-volcanic landslide in North American history when 70 million cubic metres of dirt and rock thundered down the side of the pit reducing copper production substantially. But Bingham could almost be mistaken for a natural valley. Not so for the second-deepest open hole excavated. For dramatic impact it's hard to beat the Mirny diamond mine in Siberia; although it is only 525 metres deep, its position right next to a highly populated area makes it highly dramatic.

But we don't just explore, we also exploit, and there is nothing as exciting or enticing as digging for gold. Gold is the ultimate lure. It has been that way throughout our history. Other metals can be more expensive, but they are rare. There is no metal as desirable as gold.

Another name for Johannesburg in South Africa is Egoli, or the City of Gold. It lies next to what is called the Witwatersrand arc – the richest goldfield ever discovered from which 40,000 tonnes has been excavated in the past 130 years, or, put another way, about half of all the gold ever mined. About sixty kilometres to the west of Johannesburg is Mponeng – that means 'look at me' – the deepest mine in the world.

At Mponeng every day 4,000 workers descend into an underground city and it is like working in an oven. In the depths of the mine the rock is at 60°C, too hot to touch.

Giant ducts blow cool air into the tunnels in order to try to bring the temperature down. To cool the mine more than 6,000 tonnes of ice are made every day and then mixed with salt to form a slushy liquid. This is pumped down the mine where air is blown over it. It's the only mine in the world to cool itself this way. But why is there gold here? Crush the rocks and you get a sugar cube of gold for every tonne. Why is this the richest goldfield on Earth?

The answer goes back to almost the very first years of our planet when, as we shall see, after hundreds of millions of years of cooling, it was struck by a heavy bombardment of large meteors starting around 4,100 million years ago and lasting about 300 million years. It was the final large-scale cleaning up of debris left over from the formation of the planets. Before the so-called 'late heavy bombardment' all the gold in the rocks at the surface of the Earth had percolated to the inside where it has been ever since and where it would have been forever out of man's reach. The new impacts brought new gold deposited near the Earth's surface.

At that time – called the Archean era – many believe that the Earth did not have continents, just island arcs made of volcanoes, several of which around 3,900 million years ago fused to form one of the first mini-continents called the Kaapvaal Craton which held gold delivered during the bombardment. By 3,000 million years ago rivers had eroded the rocks and carried gold-laden silt to huge deltas. Had things stayed that way it is unlikely that it would have become the richest goldfield on Earth. Water and time would have dispersed the gold still further.

The fact that Johannesburg is the city of gold is due to a giant meteor smashing into the region over a billion years after the gold deposits were first laid down. The meteor was one of the largest to have struck the Earth after the end of the Late Heavy Bombardment – between 5 and 10 km across – and it hit near the current town of Vrede-fort, making a 300-km crater. Subterranean strata were uplifted and overturned. Near Johannesburg gold-bearing strata were brought nearer to the surface while at ground zero the gold was pushed deeper underground. And so pan-handlers who flocked to the region after gold was found in a stream in 1886 were looking for bits of a meteor. The interior of the Earth and the Cosmos are connected.

A few years ago it was realised that the main gold seam at Mponeng was running out so geologists drilled explor-atory holes radiating out from the mine, hoping to find further seams of gold. They found one but it was well away from the main shafts and deeper than any current excavation. So began a project to reach it. Drilling down to a new record depth and then across took years and over 600 explosions, each advancing the tunnel two or three metres. At the base of Level 126 is a sign: 'You are now standing at the world's deepest point, 3,612 metres below the surface or 2,059 metres below sea level'. This is the deepest point underground that can be reached by humans – a third of the way through the Earth's crust – its outermost layer.

Mponeng consists of some 400 km of tunnels, many of which are abandoned. In some of them illegal visitors have set up home. They are called the 'Ghost Miners of

Mponeng' and they live a strange, furtive existence scavenging in the dark. They should not be there, but have sneaked by or bribed the lift shaft operators. They look for rocks that hold gold deposits that have been missed, extract them and even smelt the ore underground using dangerous methods that expose them to toxic chemicals such as mercury. These modern-day Morlocks stay underground for many months and their lack of exposure to light and inadequate diet turns their skin grey and their eyes sallow. Entire families live there and prostitutes regularly visit the workers. The mines' security guards occasionally round up some of them, but others have machine guns and no one wants a gunfight down there, so they are left alone.

Mponeng is one of the most profitable gold mines in the world and is yet another example of how our Earth's insides are connected to the rest of the universe, for gold, which is such a part of the Earth, has only lodged here. It comes not from our planet, but from far, far away.

On 3 June 2013, sensors on the Swift sky-monitoring satellite detected a dramatic flash of high-energy radiation coming from deep within the constellation of Leo. For a few decades now astronomers have known that from time to time flashes of intense radiation in the form of gamma rays come from deep space, far beyond our galaxy or even our local cluster or clusters of galaxies. That particular single pulse of gamma rays, about a fifth of a second long, was detected at 15:49 and 14 seconds GMT. Within seconds Swift turned its main instrument to look at it.

It was designated GRB130603B and an alert was

automatically sent out to the astronomical community for observations. Less than three hours later the huge Multiple Mirror Telescope in Arizona was turning its segmented mirrors towards it, its normal observing schedule overridden by this 'target of opportunity'. Nine hours after the blast was detected the myriad radio dishes of the Very Large Array of radio telescopes in New Mexico joined in, as later did the Hubble Space Telescope and the Gemini South telescope in central Chile, South America.

What had happened was that two dense, dead stars had collided, producing an explosion of such energy that it was seen half a universe away. The objects were neutron stars, each so dense that a teaspoon of their material would weigh five billion tonnes! When such objects smash into each other the event is detectable all over the universe. Debris is thrown out into space and it is so hot that atoms are fused together, and gold is formed, in GRB130603B's case about twenty Earth-sized planets of gold. Such events could account for almost all the gold in the universe. In our own galaxy it is thought that neutron star collisions take place every 10,000 to 100,000 years.

The gold in the Earth is a gift from the stars, from two dead stars each the remnant of a star that lived and died in a supernova explosion, scattering itself throughout space and leaving behind a compressed remnant to wander throughout space. Each time a miner, legal or ghost, in Mponeng sees a glimmer of gold he reinforces man's kinship with the stars.

But something perhaps even more remarkable than gold is found in the depths of Mponeng mine. In 2011 Tullis

Onstott and Gaetan Borgonie of Princeton University were looking at water from a rock fracture deep within the mine. To their great surprise they found roundworms about half a millimetre in length: 'It scared the life out of me when I first saw them moving, they looked like black little swirly things,' said Onstott. Bacteria have been found at such depths but no other multicellular organism has. Having recovered from their astonishment the researchers set about gathering more information about the remarkable creature which they called *Halicephalobus mephisto*, after Mephistopheles, which means 'he who loves not the light'. The nematode appeared resistant to high temperatures and reproduced asexually, requiring no mate. Its diet was bacteria, no other source of nutrition was available.

The big question is, of course, where they came from. It seems unlikely that they could have evolved in the rocks and been uncovered by the mine works. They were more likely descended from animals living on the surface that were washed down into the mine by rainwater. Nonetheless, it is remarkable that they can survive there.

From time to time some scientists speculate about a 'deep biosphere' within the earth, living creatures – bacteria – in rocks stretching many kilometres deep into the Earth. About twenty years ago, the late Thomas Gold published a remarkable paper (something he regularly did) in the *Proceedings of the National Academy of Sciences of the United States of America*, in which he postulated that a 'deep, hot biosphere' exists in the crust of the Earth, suggesting that this biosphere is as deep as 6 km. Indeed, some studies conducted about twenty years ago were quite optimistic about the extent of this new realm of life, suggesting that between 35 and 50 per cent of all life could be living in subsurface rocks. Such estimates, it now appears, were based on sparse data combined with optimism. There are colonies of life, bacteria and fungi, living in sediments below the ocean floor in the crust's deepest layers. They are widespread but not abundant or energetic and appear similar to organisms found living in oil reservoirs, so perhaps they migrated down there rather than having evolved *in situ*.

Life, it seems, needs a gradient, a change in temperature or in chemical composition, that it can use to get energy and although beneath our feet the rocks and radioactivity do provide an energy source it is not enough to support an extensive deep biosphere. Recent estimates suggest that perhaps only 1 per cent of the Earth's biomass lies way beneath our feet, and they are the least energetic 1 per cent of life on Earth.

The miners of Mponeng have dug deeper into the uppermost layer of the Earth – its crust – than anyone. But they have not come the closest to the centre of the Earth. That honour belongs to the three men who have dived to the deepest part of the ocean – the Mariana Trench in the western Pacific Ocean. It's part of the boundary between two of the great so-called tectonic plates that cover our planet, in this case one moving under the other. As the floor of the ocean dips downward a scar is created – the deepest part of the ocean, 10,911 km underwater, or about 8,000 metres deeper than the Mponeng mine.

Three manned expeditions have reached the Mariana

Trench, only two leagues in depth; 20,000 leagues is more than the whole planet. The bathyscaphe *Trieste* took US Navy Lieutenant Don Walsh and Frenchman Jacques Piccard down in 1960 and the *Deepsea Challenger* took film director James Cameron down in 2012. It's thought that the first expedition touched down in a slightly deeper area. Many years later I spoke with Don Walsh at an oceanography conference in the UK. I was amazed that he was walking around the exhibition unrecognised, yet more people had set foot upon the Moon than where he had been.

He told me some of the details about his remarkable journey and I asked him about his reaction to being, along with Piccard, the closest humans to the centre of the Earth. He was characteristically modest. 'Of course,' he replied, 'you could get closer if you dived to the bottom of the Arctic Ocean at the North Pole. Because the Earth is a flattened sphere the closest you can get to the core of the Earth is at the North Pole, it's only about 20 metres closer than the Mariana Trench, and one day I plan to go there.'

When it was time to leave Boulby mine and return to the surface there was one more thing I had to do. Kneeling, I brushed the salty dust away, revealing the bedrock, and placed my palm on the hot rock, closed my eyes and wondered. They say tactile contact with a thing is important to us if we seek to know it. Beneath me, 6,370 km away, is the centre of the Earth. For us surface dwellers that is not such a great distance. It's about the same as Paris to Delhi, or Sydney to Singapore, or about the width of the North Atlantic. But it is a journey we can never make. The Earth teases us, exposing rocks from its interior and then taking them back again. A modern-day Columbus heading straight down could only navigate his virtual journey with the aid of satellites, seismographs and supercomputers.

Down there is the history of our planet written in crystals and minerals, temperature and pressure. Below me is the divide between the Earth's crust and its major component, the mantle, and then down through the various subdivisions in the mantle, skirting vast plumes of hot rock making their way to the surface from the heat below, past descending giant slabs of sea-floor groaning and flexing as they are being crushed by the pressure, through the swirling storms in a vast sea of molten iron and then to the Earth's solid iron heart with its giant crystals of iron. All this down to 6,370 times deeper than my tiny scratch on the surface.

In the billions of years since the Hadean Epoch – the first great time period of our planet, and the quarter of a billion years since the Sun beat down on the lazy waves of the Zechstein Sea, the surface of the Earth has changed and so has its interior. Hundreds of thousands of years ago early man migrated across the continents, recently establishing civilisations, building great cities, only to see them crumble at the mercy of the little-understood forces from within the Earth. The Earth is our friend and our enemy. One day, the power of the Earth will threaten our very existence.

In my opinion the journey to the centre of the Earth is a voyage like none other we can imagine. It involves more drama, more extremes, more science and more adventure than even a journey to the edge of our galaxy and beyond. Yet we will never make that journey. We will reach the distant stars before we reach the centre of the Earth.

Like Professor Lidenbrock, I am at the start of a journey, but unlike him I will be going all the way. What started on the shores of an ancient ocean will end at the centre of the Earth. With slight sadness I climb into the lift cage for the journey back to the surface. During the ascent I noticed that my watch read 13:13.

3

Underworld

In what could almost have been a partial description of the plot of *Journey to the Centre of the Earth*, the ancient Greek philosopher Plato wrote:

> In the earth itself, all over its surface, there are many hollow regions, some deeper than our [Mediterranean] region but with a smaller expanse, some both shallower than ours and broader. All these are joined together underground by many connecting channels, some narrower, some wider, through which, from one basin to another, there flows a great volume of water – monstrous unceasing subterranean rivers of waters both hot and cold – and of fire too, great rivers of fire, and many of liquid mud, some clearer, some more turbid, like the rivers in Sicily that flow mud before the lava comes, and the lava stream itself.

To many men and for much of history the Earth was hollow with strange and significant places to visit. It was a place for the afterlife. Hell and the devil were inside the Earth as well as being the abode of the dead in many other religions. Caverns and caves were the entrance to the underworld, as in Herakleia in Pontos on the southern coast of the Black Sea and in many other places in the Greek and Roman world. There is a cave in Celtic mythology called Cruachan, in Co. Roscommon, where all the beasts of the underworld return to the surface where they once lived. Sometimes god-like men lived underground, emerging with supernatural gifts to change the course of history. In many mythologies the interior of the Earth was an alien place where no ordinary man could venture.

Eccentrics and some writers held on to ideas of a vast interior space within the Earth. It has been reported by some that the great mathematician Leonhard Euler (1707–83), who made fundamental discoveries in calculus and graph theory, devised a so-called thought experiment that had an interior sun shining on an advanced subterranean civilisation. In 1741, Ludvig Holberg's (1684–1754) novel *Niels Klim's Underground Travels* has its hero spending several years living in the interior of the Earth on the inside of a smaller shell accompanied by intelligent trees, and professes that 'the conjectures of those men are right who hold the Earth to be hollow, and that within the shell or outward crust there is another lesser globe, and another firmament adorned with lesser sun, stars, and planets'. Then there are the writings of Giacomo Casanova (yes, *the* Casanova). He produced a story of almost 2,000 pages concerning a brother and a sister who fell in love and discovered an underground utopia inhabited by the Megamicres, a race of multi-coloured, hermaphroditic dwarfs!

Strangely, it seems that an evangelical belief that the

Earth was hollow has been held by many American soldiers and thinkers. Above the grave of John Cleves Symmes Jnr (1799–1829) in Hamilton, Ohio, is a sculpture showing a hollow earth. This army officer proposed that the Earth was a series of four shells with openings at the poles. 'I declare that the earth is hollow and habitable within; containing a number of solid concentric spheres, one within the other, and that it is open at the poles twelve or sixteen degrees. I pledge my life in support of this truth, and am ready to explore the hollow, if the world will support and aid me in the undertaking.' He made a lot of money and became famous on the lecture circuit but garnered no credibility among scientists even though his ideas were expounded by many followers. He proposed an expedition to the North Pole hole and US President John Quincy Adams indicated he would support it but he left office before he could. The next President of the United States, Andrew Jackson, was less impressed.

Edward George Earle Lytton Bulwer-Lytton, 1st Baron Lytton PC (1803–73), was an English novelist, poet, playwright and politician whose words are often on our lips even if we are unaware that they are his. He coined such phrases as 'the great unwashed', 'pursuit of the almighty dollar', 'the pen is mightier than the sword' and 'dweller on the threshold', as well as that famous opening line 'It was a dark and stormy night'. One of his lesser works was *Vril: The Power of the Coming Race* (1871), a story of a subterranean race waiting to reclaim the surface of the Earth. The book popularised the Hollow Earth theory and may even have inspired Nazi mysticism. His word 'vril', which was intended to mean energy, lent its name to Bovril meat extract.

A certain Marshall Gardner wrote *A Journey to the Earth's Interior* in 1913. He actually built a working model of the hollow Earth which he patented (US Patent 1,096,102). In 1915 Vladimir Obruchev (1863–1956), a geologist and one of the first Russian science fiction authors, wrote a novel, *Plutonia*, in which the hollow Earth possessed an inner sun and was inhabited by prehistoric animals. The characterisation is poor and the plot overly simple, but the redeeming quality of the book is Obruchev's knowledge of geology. Lobsang Rampa, in his 1963 book *The Cave of the Ancients*, wrote that an underground chamber system exists beneath the Himalayas filled with ancient machinery, records and treasure. His credibility was somewhat undermined when it was discovered that he was really Cyril Hoskin (1910–81), a plumber from Devon. Lobsang Rampa went on to write many more books containing a mixture of religious and occult material. One of the books, *Living with the Lama*, was said to have been dictated to Rampa by his pet Siamese cat, Mrs Fifi Greywhiskers.

In the 1950s and for the next few decades the idea of a hollow Earth and underworld caverns went hand in hand with UFOs, with sometimes a whiff of the legend of Atlantis, but the idea's credibility, if it ever really had any, was reduced to non-scientific ramblings. That is not to say that the idea has not become a staple of modern science fiction, especially on TV. Don Rosa's 1995 story *The Universal Solvent* imagines a way to travel to the planet's core using 1950s technology. The fictional solvent has the power to condense everything except diamonds into dust. The solvent is accidentally spilled and it bores a shaft to the core of the Earth. Rosa describes in great detail the journey downwards to recover the destructive material. The prize for sheer imagination has to go to the Teenage Mutant Ninja Turtles. In season three the episode 'Turtles at the Earth's Core' has it all: a deep underground cave where dinosaurs live, and a crystal of energy that works like the Sun, keeping the dinosaurs alive.

4

Birthmarks

Every so often on their journey downwards through the Earth the three travellers, Professor Lidenbrock, Axel and Hans, stopped for a while to contemplate what they had seen. It was a literary device that Verne used to get his message across. Another way he did it was when Axel had a dream:

> Centuries passed by like days. I went back through the long series of terrestrial changes. The plants disappeared; the granite rocks softened; solid matter turned to liquid matter under the action of intense heat; water covered the surface of the globe, boiling and volatizing; steam enveloped the earth, which gradually turned into a gaseous mass, white-hot, as big and bright as the sun.
>
> In the centre of the nebula, which was fourteen hundred thousand times as large as the globe it would one day form, I was carried through interplanetary space. My body was volatized in its turn and mingled like an imponderable atom with these vast vapours tracing their flaming orbits through infinity.
>
> What a dream this was!

We live on a world of rock and metal, oceans and atmosphere. Our survival is possible only on the surface in a thin skin that does not even extend to the highest mountain peaks. We cling on to one of the solar system's inner planets – smallish rocky worlds with many similarities and yet profound differences. They are very different from the much larger gas giant worlds that dominate what many term as the central zone of the solar system before we get to the ice worlds that live in its cold and dark outer reaches. Today each world has its orbit that for the most part keeps it out of trouble. They are survivors that maintain a stability that was hard-fought. But look closely at the small rocky worlds, and if you know what to look for you will see the scars of a violent formation – birthmarks written in motions, metals and isotopes of elements rarely discussed outside of those whose passion is rocks. They tell us many things, but chief among them is that our Earth was formed with astonishing rapidity and with great violence. The Earth's story and the journey to its heart can be told in many ways, but I chose to start it 150 years ago on a May evening in the same year and the same country in which Jules Verne was just completing his *Voyage au Centre de la Terre*.

In the evening of 14 May 1864, Jules Verne was in Paris finishing working for the day on his second major book in which he had invested so much hope and expectation, and he was about to prepare for his customary early bedtime. Over southern France, however, at about 8.20 p.m., an enormous fireball sped through the sky causing considerable consternation. It was the brightest meteor for years and was described as casting shadows through lace curtains. Some people flung open doors and windows, some ran outside to see the fireball, watching it pass northwards as it lost its white-hot glow, blushed red and exploded 20 km high, leaving behind a long, thin, white smoke trail. Moments later black pieces of rock, most of them smaller than your fist, smashed into the ground near Orgueil in southern France, and the race was on to find these precious items.

Adults and children raced over the hills, some on horseback, most on foot, running through the vineyards looking for anything unusual, for this was one Easter egg hunt in which all intended to find the prize. In wagons and baskets and in the folds of aprons, the villagers collected as many pieces of rock as they could find, totalling an estimated 20 kg. The pieces were soft enough to cut with a knife and disintegrated in water. A piece of the black rock properly shaped and sharpened could be used to write and draw as though it were a stick of charcoal. They would no doubt fetch a good price when the officials from the museum arrived looking for the rock.

Jules Verne did not see the fireball but he read about it in the newspapers a few days after. Years later he wrote a book, *The Meteor Hunt*, based on a meteorite fall in 1901. It was published posthumously, having been reworked by his son. In it the Orgueil meteorite gets a mention.

What had fallen to earth that night was a very rare type of meteorite and has become one of the most famous ever discovered. It is one of only nine known to be part of the CI Chondrite group. They have been found in India, Canada, Tanzania and, remarkably, twice in France. Four have been found in recent years during a hunt in Antarctica, where meteorite spotting is easier as they show up against the ice fields. It is a special rock because, excluding the elements hydrogen and helium, it has the same composition as the Sun and hence the gas cloud out of which our Sun formed. Pieces of it are in many museums around the world. The Orgueil rocks contain secrets about the Sun's and our Earth's formation out of a vast placental cloud of gas and dust that was once drifting between the stars.

We will return to the Orgueil rock. To journey to the centre of the Earth we must first understand where it came from and why it is like it is. Only then can we descend, know the way and recognise the signs we are seeing, why they are there and what they mean, for inside the Earth is its past and our future.

We begin our journey to the heart of our planet far away in time and space with a distant star that lived and died long before our Sun and the Earth was born. When the universe was young there were only the simplest elements, hydrogen and helium – products of the Big Bang – and although you can make stars out of them you can't make planets, or at least not proper ones. Planets need heavier elements, oxygen, silicon, magnesium and sulphur, for example. Stars are the chemical factories in which these elements were assembled and when they exploded as supernovae they enriched the cosmos, allowing subsequent generations of stars like our Sun to be born – stars capable of forming rocky planets. Our Sun's ancestors were the first stars to be born after the starless epochs that followed the Big Bang. Astrophysicists believe they were larger and brighter than our Sun. Inside their cores high temperatures fused the hydrogen and helium together to make heavier elements. Fortunately for us many of these stars were unstable, and at the end of their lives exploded, scattering their enriched material throughout space. Indeed, in the high-temperature eject thrown from these supernova, further so-called nuclear synthesis took place creating rare, short-lived radioactive elements that have proven useful in determining the chronology of our planet's birth, for it was from this stellar wreckage that our Sun, and the Earth, were born.

As the first generation of stars was dying and paving the way for the development of our own galaxy, the gas ejected from these objects sometimes gathered into a huge cloud that formed a dark silhouette against the stars. Long ago the atoms that were to make the Sun were in such a cloud, as well as those destined to form the Earth and the other planets, and those in you and me and in all the creatures that have ever lived. Then they were drifting across hundreds and hundreds of billions of kilometres in almost empty space, far from the light of stars. This molecular cloud, as astronomers call it, moved slowly, rotated slowly. Molecular clouds are still around today and we see them as the largest inhabitants of galaxies, up to 300 light years across. But as for our cloud, the eons passed and it followed the stars and other such clouds in orbit around the centre of our galaxy. Thus gas and a collection of minuscule grains formed in previous stellar atmospheres floated

around in a molecular cloud at a few degrees above abso-
lute zero, in what we would today regard as a near-perfect
vacuum.

But the cloud was on a knife edge, teetering on the verge
of collapse. If it had a mass above tens of thousands of
times the mass of our Sun, as we think it had, it became
unstable and fragmented, eventually becoming an open
cluster of hundreds or thousands of stars. Smaller clouds
have different fates. A nearby supernova can trigger their
collapse. The material from the exploding star would
scatter most of the cloud but computer simulations reveal
that the central part of the cloud would be compressed a
millionfold, enough to cause its collapse and initiate the
process of star formation.

At the centre of the infall the embryo of a star is formed
and its gravity pulls more and more material on to it. It
becomes hotter, especially at its core. But not all the material is captured. Some ends up in orbit of this protostar
in what is called a circumstellar disc, and the gas near the
star is hotter than the gas further out. This temperature
gradient is one of the most important factors in the formation of planets. The gas around the young star is mostly
hydrogen and helium but the small proportion of heavier
elements start to condense out of the solar nebula, forming tiny grains that then start to coalesce into larger ones.

Initially the composition of the small bodies that formed
in the solar gas cloud reflected the position at which
they formed. Close in, where it was hotter, it allowed
only heavier elements to condense. Moving further out,
lighter elements condense from the gas until the 'ice line'

is reached where water condenses and the growing fragments can be covered in ice. Only when you get close to
the ice line do you expect to find iron in the region where
water reacts with iron to get iron oxide, which is an important component of the Earth. As far as we understand,
the body that was growing to become the early Earth did
not initially contain many volatiles or light elements.

All this happens in secrecy, for the protostar and its circumstellar disc is still within the larger gas cloud. But that
is soon to change. Dan Frost from the University of Bayreuth in Germany has made a particular study of these
earliest times in the history of our planet: 'In the beginning
everything is gas and that gas starts to condense to a dust
and the dust starts to accumulate into little marble-sized
things and gravity works to make them larger and larger.'
Soon this accretion phase ended as the young Sun enters
what is termed the T Tauri phase, named after the first
star in which it was observed. During this time the Sun
would have a strong outflowing solar wind that dispersed
the gas in the solar nebula, leaving only the small rocky
bodies with their various compositions depending upon
their distance from the protosun. It is about 4,568 million
years ago.

'Larger and larger the objects gathered as gravity has
its inexorable way,' adds Dan Frost; 'when they reach a
certain size – thought to be about 500 metres – the pressure in their interiors and the elevated temperature inside
starts to change them. Their insides melt and the components of the rocky material separate. The energy from the
radioactive decay of short-lived elements – made in the

cauldron of the ejecta of the death of previous generations of stars – such as Aluminium 26 – cause additional
internal heating.' Inside these small rocky bodies cores
were being formed, they were changing from mere chunks
of rock. Metal, mainly iron, was accumulating at their
hearts. Only three million years after the process started
the building blocks of planets were everywhere, and one
of them has survived to the present day.

In July 2011 Nasa's *Dawn* spacecraft was far from
Earth and approaching the asteroid Vesta, which, with
a mean diameter of 525 km, is the second-most massive body in the asteroid belt – a region of rocky bodies
swarming between Mars and Jupiter – objects taken to be
leftovers from the formation of the planets and some that
were going to form planets but were prevented due to the
gravitational disruption of mighty Jupiter. *Dawn* had been
in space since September 2007 and was the first spacecraft
to visit Vesta. It had acquired its first targeting image three
months earlier when it was over a million kilometres away
and Vesta barely more than a splash of light. A month
later it used its ion thrusters to slow itself down slightly
so that it was travelling at the correct velocity for Vesta
orbital insertion. Much to the relief of scientists all over
the world the manoeuvre was successful and *Dawn* immediately fired its thrusters again to spiral down to enter
a survey orbit taking sixty-nine hours to circle Vesta at an
altitude of 2,750 km. Later it moved much closer to carry
out mapping from a 680-km orbit. It was late summer in
Vesta's southern hemisphere and a large crater dominated
the asteroid's South Polar regions.

Vesta is a remarkable small world because it offers us a glimpse of what the solar system was like when it was young. It could be the only remaining example left in our solar system of a planetesimal, one of the solar system's building blocks that came together to form the rocky planets like our Earth. Far beneath the craters and gullies of its surface it has an iron core – perhaps 220 km across – formed when its interior became hot enough to melt. Today Vesta is almost alone but once it was one of many millions living and dying in violent collisions as they crowded our young Sun. Vesta is a chance survivor, a frozen snapshot of a process that has run to completion everywhere else.

Initially the proto-Earth did not have much in the way of light elements because they could not condense out of the gas because the Sun was too hot at that distance. But things were to change. The planetesimals collided with each other sending fragments in all directions, mixing up the previously smooth gradient of composition radiating outward from the Sun. In this way the Earth was enriched with material that had condensed in the cooler regions further away from the Sun. Dramatic things were happening in the outer reaches of the solar system. The giant planets Jupiter and Saturn had formed very quickly and were now shifting their orbits, travelling outwards and throwing a lot of volatile-rich planetesimals towards the inner solar system and the proto-Earth.

It all happened remarkably quickly. The process of transforming nebular dust into fully formed planets took less than a hundred million years out of the 4.6 billion years of our solar system's history. The formation of planetary cores was, however, much swifter, only about a million years after the first solids condensed. The primitive Earth was perhaps about half the size it is now, heavier elements draining to its core and the lighter elements rising to the surface as it was being bombarded. Despite these punishing impacts and major internal changes the Earth was on its way to becoming a planet we would recognise. The outer layers of Earth were made of lighter silicate rocks formed when the molten surface cooled and there may even have been oceans of steaming water in the periods between the catastrophic impacts that vaporised them and turned the surface back into hot magma. Astronomers watch the scene regularly. In computers researchers scatter rocky bodies of various sizes in various orbits in the early solar system and observe as they collide. Eventually, the small worlds are replaced by a few larger ones, and soon afterwards the impacts between them become truly cataclysmic.

The Earth settled down very quickly and by 4,417 million years ago its surface had cooled so that liquid water existed on its surface and interior temperatures were not that different from what they are today, indicating that in the interim the Earth has cooled, but not by that much. It is even possible that the first stages of life may have been taking place in the short-lived oceans of Earth. Looking at samples of the world's earliest rocks, scientists cannot help but be impressed by the swiftness at which primitive life started on our planet. So there we have it. The Earth had been formed, although it was somewhat smaller than it is today. It had a noxious atmosphere but liquid water on its surface. A primitive crust had formed possibly from the outwelling material from volcanoes and inside it had a core and an overlying region of rock we call the mantle. It was in many ways the planet we know today.

Curiously, recent work may indicate that as much as half of the water in the Earth's oceans could be older than the Sun. Detailed computer simulations of the behaviour of the dust and gas cloud suggest that the Earth may have inherited water from the cloud instead of it being formed later. It had been thought that the ice crystals in the cloud that would have formed as it condensed would have been vaporised and turned into hydrogen and oxygen, but now scientists are not so sure. If water had to re-form later on then some speculated that the way it did it might be specific to our solar system, and thus many planetary systems formed elsewhere in space might not have as much water as our system does. Since water is essential for life as we know it this might mean that other planetary systems might not be suitable for life. It now seems that water is a universal ingredient for planetary formation, and, taken with evidence that planets are common in the universe, suggests that planets like the Earth, with the possibility for the development of life, might be widespread.

Then, 4,445 million years ago it happened.

If you had been standing on the primitive Earth just over 4,400 million years ago, on an island perhaps, up to your final hour it would be difficult to make out the motion of the object racing through space towards you, but each time you looked back it would appear a little larger – this planet the size of present-day Mars. In the last twenty minutes its incoming motion would become frighteningly obvious and the outcome understood as inevitable. It is going to make an off-centre impact and there is nowhere you can go for safety. If there are any primitive stirrings of life in those ancient oceans they are about to be snuffed out. When it arrives, the impactor cuts through the atmosphere in seconds, forcing it out of its way in supersonic winds, filling the sky with sonic booms for a split second. Tidal forces would have caused the Earth and the impactor to be slightly pear-shaped as they approached, and there must have been an instant when the two worlds were just touching, filling each other's skies with rock, wiping out the skies of these two separate worlds for ever.

Instantly hundreds of cubic kilometres of rock are flash-vaporised and blasted into space as the two worlds grind themselves together. The expanding line between the worlds quickly encompasses them both in a white-hot embrace. In minutes a major part of the Earth is missing and the rest is turning an ever-brightening red. Streamers of superheated rock vapour stream into space and shock waves shudder in what is left of the two worlds. Our new Earth and the Moon were born and nothing was ever the same again. For a while the two wrecked worlds are connected by a thin bridge of glowing material that starts to fall on to one or the other or fragment into a string of beads. The distorted impactor staggers to a maximum distance of a few Earth diameters and then falls back a second time. This time it is fully destroyed. The Earth and its new material is now fully molten – a fluid world rippling slowly under the bombardments. Within hours of the impact a ring of debris forms around the Earth, which is still far from spherical. Of all the material thrown into space after the collision, most, but not all, of it is eventually drawn back to Earth. The fraction that remains in space became our Moon. Slowly the bombardment ceases, and the Earth can evolve.

Remarkably, the Earth recovered quickly from its period of hell and its surface solidified, forming a hard skin by about 4,400 million years ago, a skin so resistant to weathering that it is still there today. The magma ocean crystallised from the bottom to form the mantle. The very first rocks are gone – all that remains is splinters, if you know where to find them.

5

The Survivors

The Jack Hills are an isolated region in an almost empty part of one of the world's largest and emptiest countries. They are a low range in the Murchison District of Western Australia. In some respects the region's remoteness is a blessing for different types of scientists. A few hundred kilometres away a major international astronomy project is under construction. It's a series of hundreds of small radio telescopes called the Square Kilometre Array. Observing the universe in radio wavelengths' size and sensitivity will help it make many new discoveries. But it has to be kept away from man-made radio interference and this part of Western Australia is ideal. It's sheep farm territory where people live in homesteads not villages. There are few roads and scientists travel overland, parking their 4x4s and setting off up the Jack Hills looking for one particular rock outcrop.

Simon Wilde's interest in geology goes back his schooldays, growing up in the shadow of the Wrekin, or, as he came to call them, the Uriconian (Neoproterozoic) hills that stretch across Shropshire in the Welsh Borders of Great Britain. After studying geology he joined the Western

Australian Geological Survey and was given the task of mapping the rocks near Perth. He later said, 'Was there any better job in the world at that time?' It was a job that would lead him to discover the oldest rock ever found.

He first heard of the Jack Hills in Western Australia while preparing a research report for the Second International Archaean Symposium held in Perth in 1980. He moved to the Western Australian Institute of Technology (now Curtin University) and, along with fellow researcher Bob Pidgeon, proposed an investigation of the rocks along the western margin of the Yilgarn Craton – a 2.8-billion-year-old granite and metamorphic structure. 'We were successful – and so begins the story, since one of these was Jack Hills!'

In the next few years Wilde and a team from the Western Australian Institute of Technology, along with John Baxter and Bob Pidgeon, drove some 800 km north from Perth to the Jack Hills. It was clearly a region with some ancient rocks. At the time the nearby Mt Narryer, 60 km to the south-west, had yielded the oldest crystals on Earth. But little was known of the geology of the Jack Hills.

The Earth's first outer layer – the crust – was almost pummelled into oblivion in the tens of millions of years following the formation of the Moon. A few hundred million years later the Late Heavy Bombardment again tortured it when a torrent of rocky debris, ranging in size from today's continents downwards, spent tens of millions of years reworking the surface. But, shattered and splintered, the crust survived as tiny grains so tough that they lasted over four billion years and have been incorporated into almost every form of rock since. Look hard enough and you will find them in rocks laid down under seas, spilled out by volcanoes or rocks transformed by heat and pressure. They are found among the grains of sand on beaches and in deserts, among the frozen soil of the tundra and in the rich soil of the rainforest. They are tiny and almost everywhere. Scientists call them zircons.

'We got a rather poor yield from several of the samples so we concentrated on sample W74, thinking that this was the most likely to contain the oldest grains. Zircons from this sample were duly mounted in epoxy and taken to Canberra for analysis. I was later to receive a telephone call from Bob Pidgeon saying they had hit the jackpot — two grains some ninety million years older than anything so far obtained from Mt Narryer.' The processed age was 4,276 billion years.

The Jack Hills terrain was mapped over several years but no older material was found and, by the mid-1990s, interest in the area had started to wane. It took a novel idea for research into the earliest splinters of rocks to get researchers back to the Jack Hills.

Among those researchers is John Valley, the Charles R. Van Hise Professor of Geoscience at the University of Wisconsin. He has always been interested in rocks. 'We lived in Boston when I was young and my father took me on a geological trip and I've had a sharp-pointed hammer and a rock collection ever since.' He is particularly interested in metamorphic petrology, which is the recrystallisation of rocks under high pressure and temperature, and in particular the chemical and mineralogical changes that occur in such extreme environments. He has made several trips to the Jack Hills. 'It's an amazing place. At about five or six hundred kilometres north of Perth and Perth is already an isolated city. Fifty kilometres before you get to Jack Hills you pass the last inhabited station. It's all just amazingly flat. There are these low hills that run in a north, east–south, west trend. The reason they are hills is because they are these very hard quartz rock conglomerates and sandstones that have resisted erosion. When we first went we saw a little bump on the side of the hill and decided to look at that.' Bob Pidgeon and John Valley collected a total of thirty-one samples from a rather prominent outcrop on the eastern flank of Eranondoo Hill. Little did they know that sample W74 was to provide the oldest rock fragments ever discovered on Earth.

John Valley: 'About fifteen years ago I had a student and we were looking at rocks of different ages. This was William Pack, who is now a professor at Colgate University. We were analysing oxygen in the mineral zircon of different ages. One of the things we thought was that we should analyse the oldest oxygen we can find from the Earth that might have been trapped in minute quantities inside the zircons.' It would be a glimpse, they believed, of the Earth's primordial composition.

Valley first tried to get zircons from Canadian rocks that are four billion years old but there was a lot of radiation damage that made measurements unreliable. 'Then, at a meeting in 1998 in Beijing, I met Simon White and we got talking. Simon had been involved in the discovery of 4.2 billion-year-old zircons from the Jack Hills and I asked if I could measure the oxygen isotope ratio in them. He said he didn't have the exact 4.2 billion-year-old zircon that had been published about ten years previously, but he did have a bottle of undated zircons from the same sample and he would date some more and see if he could find me a very old one. He set about doing that in 1999 and immediately found a zircon that was 4.3 billion years old on the second day of analysis which at that time was older than anything that had been found before. Even the 4.2-billion-year zircons were shockingly old when they were discovered because everybody thought that the early Earth was so violent that nothing could survive it. And so when he found a 4.3-billion-year-old sample he got really excited. It was a little bit scary because it countered what everyone thought.'

Simon White sent the crystals to John Valley and William Pack, who took them to the University of Edinburgh, which at that time had the best equipment in the world for measuring oxygen isotope ratios. 'The problem was that the zircons were so tiny that they couldn't be analysed by conventional means. In the laboratory at the University of Wisconsin I can analyse samples about milligram size, but the equipment in Edinburgh enabled me to analyse samples a million times smaller. We published our results in the journal *Nature* in 2001. Unknown to us, another group heard of our discovery and went to the Jack Hills and worked on their samples at the same time and published their paper back to back with ours with similar results, although they didn't have a zircon that was quite as old as ours.'

Because there were two papers from two independent laboratories and groups that were reaching very similar conclusions, while there was surprise that such old samples of the Earth had been found there was also widespread acceptance. 'To my knowledge,' says John Valley, 'there are six published zircons – two of them ours – that have ages older than 4.35 billion years.' The W74 zircon was part of the Earth's original crust solidified when the Earth was in its infancy and a very different place presided over by a new-born Moon.

But as well as a record breaker the zircon was a puzzle. The oxygen isotope data in the ancient zircon suggested it had been part of a crust that had previously interacted with relatively cool surface waters. How could this be? The Earth at that time was, well . . . hell, and any surface water would have been turned instantly into steam. But the evidence kept coming. This particular zircon had seen water. A companion oxygen isotope study that was initiated in 1999, involving Bob Pidgeon at Curtin and Mark Harrison and Steve Mojzsis from UCLA, based on zircons obtained from another sample that was collected independently from the same site as W74, reached a similar conclusion. Data from the Eranondoo zircons was rewriting the earliest history of the Earth.

For millions of years after the titanic collision that formed the Moon, the Earth's surface was molten, turbulent, hissing and spitting boiling rock. Its atmosphere was steam, sulphur and other noxious gases, so it is no surprise that scientists call this epoch the Hadean Epoch, because it conjures up visions of Hades. Whatever primitive forms of life had existed before the collision were destroyed and the Earth was lifeless once again. But with hardly any atmosphere the surface radiated away its heat and as it cooled a thin crust started to form. Physics tell us that it is quite difficult to keep a rock red-hot at the surface of the Earth for very long. It will crust over in a couple of million years. A red-hot rock surface radiates heat to space about 100,000 times faster than if you had cooler solid rock. You just can't keep magma oceans on the surface of the Earth for hundreds of millions of years.

The thing that zircons tell us is that by 4.4 billion years a crust had formed that was chemically different from the rocks below it. Some call it a proto-continental crust. We don't know if there were full continents since we don't know how much of this kind of rock existed. Some say that there was a full continental mass with the same masses as today, but we don't really know that for sure.

There is another factor that determined the conditions on the early Earth and its first crust. Astrophysicists believe that the Sun at that time would have been fainter than it is today. Once we accept what the geophysicists and the zircons are telling us about the early Earth then the question of the young Sun becomes really important. The luminosity of the Sun between about 4.4 and 3.5 billion years ago should only have been about 70 per cent of what it is today. The majority of energy reaching the surface of the Earth, if it is not magma, comes from the Sun. So the question is not why was the early Earth like hell, but why wasn't it like a snowball? The Hadean Earth could have been covered by glaciers all the way to the Equator.

John Valley believes these zircons formed from remelted rock that had once been on the surface of the Earth and that had interacted with liquid water, and it required a low-temperature interaction to produce the results in the zircon we were analysing. The oldest rocks we have, samples of the Earth's first crust, are these tiny grains no bigger than a human hair. It's exactly as the poet William Blake wrote, Valley says: 'to see a world in a grain of sand'.

So I glance down a microscope at a zircon crystal. It's mounted on a glass slide and, holding it up to the light, it is barely visible to the naked eye, but through the microscope it becomes a world of its own – a tiny crystal itself containing smaller crystals. As I adjust the focus I can home-in on different facets of the crystal each brightening momentarily as the focus point passes through them. It has discolorations caused by minerals trapped inside ever since it was formed countless millions of years ago. I linger at the eyepiece contemplating the oldest surviving piece of planet Earth.

But there are rocks only a little younger than the significant zircons that you can hold in your hands, and they are found only in one place on Earth.

At the beginning of the pioneering silent film *Nanook of the North*, made in 1922, we read the words: 'The mysterious Barren Lands – desolate, boulder-strewn, wind-swept – illimitable spaces which top the world.' It is to these barren lands that you must travel to hold in your hands the oldest rocks in the world. Rocks still on the surface after more than four billion years and formed when the Earth was a very different place from what it is today and not long after the impact that made the Moon. If the life of the Earth was represented by a day then mankind appeared just twenty seconds before midnight. These rocks have been waiting since about 1 a.m.

To reach them you must take a thousand-Canadian-dollar flight of five hours and ten minutes via Air Inuit from Montreal travelling almost due north in a fifty-seater Bombardier Dash. Your third stop is the town of Inukjuak, close to where *Nanook of the North* was filmed, on the shores of Hudson Bay in Canada. Look at any series of maps or animation showing how the continents have moved over hundreds of millions of years and you will always be able to pick out Hudson Bay. It's a long-lived feature that has been relatively unchanged for eons. Inukjuak has a population of about 1,500 and makes a living from tourism. It has one hotel and no restaurant.

Inukjuak is located on the north bank of the Innuksuak River, known for its rapids and its turquoise waters. The many archaeological sites scattered along the river are evidence of thousands of years of inhabitation. Despite everything, man has clung on here. In summer the land is one of gently rolling hills, open spaces and drab outcrops of greyish brown rock. At the beginning of the twentieth century, the area was given the name Port Harrison and the French fur-trading company Révillon Frères established a post here. The Hudson's Bay Company opened its post in 1920 and in 1936 they bought out Révillon Frères. A post office and a Royal Canadian Mounted

Police attachment opened in 1935, a nursing station in 1947 and a school in 1951. In 1962, the co-operative store opened and, in 1980, Inukjuak was legally established as a municipality. But the people were slow in coming. While civilisation arrived most Inuit continued their traditional nomadic lifestyle and only began settling in the village in the 1950s.

Forty kilometres south, an hour and a half's boat journey along the shoreline, is the Nuvvuagittuq greenstone belt. The stone is hard and fine-grained, and from this geologists know it cooled rapidly. It feels like many other rocks you find in the world's oldest places but when holding this particular rock one should bear in mind that it is the only place on Earth that has rocks known to have survived from the Hadean Epoch, the first geological epoch after the Earth. The Earth was cooling after hundreds of millions of years of impacts; it had a poisonous atmosphere and lightning continuously crackled. Magma oceans simmered. We have almost nothing from between 4.4 and 3.8 billion years ago.

Jonathan O'Neil of the University of Ottawa has always been interested in the early Earth and how the first continents formed. Studying geology he asked questions for which he received no satisfactory answers so he looked deeper and became fascinated by rocks – the oldest rocks – and was drawn to the strange Nuvvuagittuq greenstone belt. 'They are very unusual-looking rocks. It has nothing to do with the fact that they are old that is drawing your attention. Usually old rocks are typically very dark or black but these colours are really unusual. Light beige.

We never saw this kind of rock before. What interested me is that these rocks record in their chemistry the conditions in which they were formed.' O'Neil took the rocks back to his lab to torture them – his words. He powders them or cuts them into slices so thin that the light can shine through them, revealing their inner structure. Initially he thought that the Nuvvuagittuq greenstone belt rocks would be like the oldest rocks then known which were from Greenland at 3.8 billion years old. But he soon realised that these rocks were much older. They had been changed since they were formed but they still contained isotope clocks. Different versions of the same element – isotopes – change at different rates. Measure them, such as the elements Samarium and Neodymium, and you can determine the age of the rock. Not only were they old, they were strange. They had formed in a shallow sea. Jonathan O'Neil says, 'The image we have is that the Hadean was a big ball of fire and that the Earth was all molten. That's wrong. It's changing a lot now.'

Judith Coggon, a young scientist at the University of Bonn, expresses the current feeling about these new views of the early Earth particularly well. 'It is definitely becoming more clear what the early Earth was like and, actually, the most exciting thing that's happening at the moment is that we're finding more and more evidence that the Earth became similar to what we see now very quickly after it was formed.'

But how have the Nuvvuagittuq rocks survived for over four billion years and resisted being subsumed into the Earth? It used to be thought that such long-lived continental structures called cratons lasted because they were made of light rock that floats on the denser rocks beneath. Then it was discovered that such structures have deep roots that penetrate the hot, semi-plastic rocks below, providing some form of stability, but in truth we don't have a complete explanation.

Jonathan O'Neil pictures himself standing next to the Nuvvuagittuq rocks, but not today, as he has done many times, but soon after they were formed, 4,406 million years ago. 'These rocks formed pretty much at the same time as the Moon – or right after the Moon was formed. You look up at the Moon from these rocks and you could probably see volcanoes on the Moon at that time.'

Knowing what the early Earth was like is vital to understand our forthcoming journey, for the secrets of the Earth's structure were hidden during its earliest times. In the billions of years since the Hadean Epoch the surface of the Earth has changed, but its interior, with one great exception, is broadly the same. Early man migrated across the continents, established civilisations and built great cities, only to see them at the mercy of the little-understood forces from within the Earth. We have witnessed the birth of our planet. Now, before we descend, we have to map our route, and for that we need earthquakes.

6

The Messengers

The start of the day on 18 April 1906 was a routine one in
the pride of the west, San Francisco, a bustling city of half
a million, but at thirteen minutes past five in the morning
its peace was shattered. A minor tremor shook the city,
lasted for sixty-five seconds and then stopped. People
breathed a sigh of relief. Perhaps that was it. Then, ten
seconds later, the big one hit; this time it lasted for two
and a half minutes. One man out early later said he saw
the earthquake coming towards him 'like the waves of the
sea'. Kathryn Hulme, the novelist, wrote, 'A far grumble
of earth roared beneath the house under our feet. A crack
opened in our wall and spread like a vein down to the slat
beneath the plaster. We fled down the hall past lurching
pictures and stopped in terror as some Alaskan harpoons
fell from their antler prongs . . .' In hotels people rushed
down from the upper floors in their nightwear. The streets
filled with people, some cried: 'Repent, repent, the Lord
has sent it.'

Gas mains were ruptured, and within moments the inferno had begun. For a few hours telegraph offices relayed
the calamity to the wider world, before the flames reached

them and the stricken city was cut off. The wooden shanty-town had fallen at the first tremor. Fire wagons connected their hoses to the fire hydrants but no water came. The city was doomed.

Throughout history cities have been wrecked by earth-quakes, villages overwhelmed by earthquake-induced landslides and coasts scoured by tsunamis. Lisbon in 1755 and San Francisco in 1906: both were turning points. Lisbon for dispelling myths about the causes of earth-quakes, San Francisco for the building and emergency regulations that were subsequently brought in. Lisbon was also a thriving city. One of the busiest ports in Europe, the wealth of its merchants was legendary. It had experienced earthquakes before, but none in recorded history like that of 1755.

It was All Saints' Day and people were on their way to church. At 9.40 a.m. a 'strange, frightful noise under-ground, resembling the hollow distant rumbling of thunder', was heard. It lasted three minutes. People ran into the streets and shouted to God for mercy. There was a pause, and the hope once again that it had stopped. Then a second shock. Dust enveloped the city and the cry was heard: 'the sea is coming.' It did in three devastating tsunamis.

Titians, Correggios and Rubens were burned in the re-sultant fire. Women in fine dresses ran frightened alongside paupers scampering over rubble and the dead to outpace the tidal wave, but none could move fast enough as it swept away all in its path. Animals ran free, and after the tsunami seemed to subside looters headed for the Mint. A pause; or was it the end? Priests moved among the dying, administering last rites. Many died unconfessed. Two hours later a third shock finished what remained of the city and the hopes of those struggling free from the wreckage and the fire.

Madrid, 180 km away, felt the shocks of what was the worst disaster recorded up to that time. Damage in North Africa was as great as that in Lisbon. Scotland, Norway, Sweden and America experienced abnormal tides. But for some the religious shock was as intense as the shaking of the ground. The King of France promised to receive the sacrament at Easter, and to give up his mistress. What was to blame? Was it the burning of heretics? In England John Wesley wrote a small bestselling pamphlet 'Serious Thoughts occasioned by the late Earthquake in Lisbon'.

The Lisbon earthquake of 1755 marked a turning point in our understanding of the Earth. It was the first quake to be studied scientifically with observations collated and scrutinised and used to create a picture of what happened. It made people think and change their view of such natural catastrophes. The Reverend John Michell was the professor of geology at Cambridge and he declared that earthquakes definitely originated from within the Earth. 'Earthquakes were waves set up by the shifting masses of rocks miles below the surface . . . the motion of the earth in earthquakes is partly tremulous and partly propagated by waves which succeed each other.' Only three years earlier the Royal Society had said that 'earthquakes only occur when people need chastising'.

Overall the Earth does not release much of its energy

in earthquakes. Our planet is slowly cooling following its energetic birth. But not cooling all that much, because its hot interior is reasonably well insulated by its outer mantle of rock. Some 46 terawatts reaches the surface from the inside, some from the core, most from the mantle, the cooling of our planet being offset by internal heating due to the decay of radioactive elements. This energy is enough to drive the internal and surface motions of our planet but it is far less than the energy the atmosphere and the surface receives from the Sun. Each year only about 1 per cent of the energy coming from below us is released in the form of earthquakes, but even that, by human standards, is a lot of energy.

Every day, across the Earth, there are fifty earthquakes strong enough to be felt and every few days one strong enough to cause structural damage to buildings. It is estimated that there are 900,000 quakes a year, mostly too small to be felt. But each year on average there are twenty quakes that cause serious damage and every few years a great one strikes. Earthquakes occur in the skin of the Earth – its crust – where rocks are cool and brittle. Strain and tension builds and when it exceeds the rock's strength an earthquake takes place. Earthquakes have told us more about the structure of the Earth than anything else. The shock waves they send out, pushing and shaking the Earth, traverse the globe and delve deep into its interior reflecting and refracting off the structures it finds there. These shocks, or seismic waves as they are called, are the messengers from the deep.

China has always been plagued by great earthquakes.

The one of 1556 killed about 800,000 people and so it has been throughout their history. But one day the court astronomer of the Han Dynasty decided to do something about it and he constructed a device to monitor earthquakes. A replica of it now stands in the Exhibition Hall of the Museum of Chinese History in Beijing. It is a large urn called the Houfeng Didong Yi, which means 'Instrument for inquiring into the wind and the shaking of the earth'. There are eight sculpted dragons crawling down its outside and each has below it a toad with an upturned open mouth. It is a replica of a device said to have been invented by Zhang Heng (AD 78–139) in AD 132. Heng was a brilliant mathematician and nowadays we would call him a scientist. Like the events in the sky the Chinese believed that large-scale natural occurrences were signs from heaven, judgements on the empire and emperor, where his mandate of heaven could be reaffirmed or rescinded.

Heng's machine was set up in Jing Shi (present-day Luoyang in Henan Province), the national capital. On one occasion, or so legend has it, a ball fell from a dragon; when no tremor could be felt, scholars were mystified. But many days later came reports of an earthquake in Longxi (present-day Western Gansu Province) a thousand kilometres away. We do not know exactly what the working mechanism was inside the device. Seismologists of the nineteenth and twentieth centuries have speculated on the mechanisms that would duplicate the behaviour of Chang Heng's seismoscope. The idea is that a faint tremor from a distant quake will cause the pendulum to swing in its direction. The position of the ball that had fallen gives the direction of the quake. Legend has it that it once detected an earthquake and five days later a rider arrived with confirmation that it had taken place in the direction indicated by the device. It is a good story but I don't think it is true, as the device would probably have been ineffective. Chinese writings talked about Chang Heng's seismograph and ones like it until the Mongols overran China. After that, it vanished as though it had never been. The next seismograph was invented in France in 1703.

Jean de Hautefeuille (1647–1724), a man full of brilliant ideas but who never stuck with them long enough to make them work, proposed filling a bowl to the brim with mercury, so that an earthquake would cause some of the silver liquid to spill out. In order to determine the direction of the shock, the mercury spilling out in each of the eight principal directions of the compass was to be collected in cavities or other containers, much like the toads in Heng's device. Hautefeuille believed that instruments would make an important contribution to understanding the shaking of the Earth and said a large number of instrumental observations would be needed to learn about earthquakes. Unfortunately it may have been one of his abandoned ideas, for there is no firm evidence that his device was ever built.

A short while later, in 1731, Italian inventor Nicholas Cirillo used simple pendulums to study a series of earthquakes that struck in Naples over a period of twenty years. Then Andrea Bina (1724–92), a teacher of philosophy at Benedictine monasteries in Italy, proposed using a pendulum with a pointer attached to its lower end, above a tray

61

of fine sand. The relative motion of the pendulum and the Earth was to be traced in the sand to see if it was 'regular or swaying, tremulous or irregular'. Bina seems to have built his instrument, but we do not know if an earthquake was ever recorded with the device.

The technology of detecting earthquakes, particularly distant ones, took a long while to develop and it wasn't until the late 1800s that seismometers became in any way effective. But it was remarkable what could be done with simple observation. In 1764 Sir William Hamilton was appointed as Envoy Extraordinary at the court of the King of Naples. A man of wide scientific interests, he spent much time studying the volcanic outcrops, sending back salts and sulphur samples. He saw the eruptions of Vesuvius in 1776 and 1777 as well as the intense earthquake that struck Calabria in 1783. He believed that earthquakes were triggered by volcanoes and wrote, 'present earthquakes are occasioned by the operation of a volcano, the seat of which seems to be deep'. He sent a report to the Royal Society about the damage caused by an earthquake over the stricken area, pointing out that the greatest damage was not at the estimated source on the Earth but some 35 km from it. 'I plainly observed a graduation in the damage done to the buildings, as also in the degree of mortality, in proportion as the countries were more or less distant from the supposed centre of the evil.'

But Sir William Hamilton – the volcano lover – is known in history for other reasons than his contributions to the study of volcanoes and earthquakes, and this is a pity. His first wife had died in 1782 and in 1783 he was back in

England, visiting his estates in Wales and his nephew's estates in Scotland. In London he met Emma Hart, who was then his nephew's mistress. Hamilton and Emma married five years later. The ceremony was a quiet one in St Marylebone parish church on 6 September 1791, two days before the couple's return to Naples. Emma signed the register as Amy Lyons. His wife became Emma, Lady Hamilton, and the story of the mutual infatuation that started when Nelson visited them is well known.

While much was being learnt from visual observations of earthquake damage the search continued for ways to record and measure earthquakes. The construction of instruments in eighteenth-century Italy frequently coincided with periods of unusually high local earthquake activity. The Calabrian earthquakes of 1783 were responsible for the jump in interest. People living in areas affected by these shocks used liquid-filled bowls and delicately balanced objects as seismoscopes. The shocks also spurred the invention of more sophisticated instruments with varying degrees of effectiveness. In 1783 Domenico Salsano, a Neapolitan clockmaker, invented a 'geosismometro' that seems to have been working shortly after the first large Calabrian earthquake. It was a common pendulum, eight and a half 'parisian' feet long. The pendulum mass was equipped with a brush, which was to record the motion of the mass with slow-drying ink on an ivory slab. Also interested in the Calabria earthquakes was physician Domenico Pignataro (1735–1802), who made an attempt at a scale for earthquake severity. He had five categories: slight, moderate, strong, very strong and violent.

In 1839, following some minor earthquakes in Scotland, James Forbes invented an instrument in which the pendulum was turned upside down to give a rod holding a metal ball supported vertically by a stiff wire. In essence the ball remained stationary while the Earth moved around it. It seemed to offer certain advantages in that a large pendulum was no longer needed to obtain longer periods of swing. It was probably the first seismometer deserving of its name.

But what were these crude seismometers measuring exactly? One man decided to find out. In the mid-nineteenth century an Irish engineer buried a keg of gunpowder on Killiney Beach, south of Dublin, laid out a long fuse, ignited it and ran to a safe distance before ducking behind a barricade. The gunpowder exploded and, using a primitive form of seismometer half a mile away, the engineer measured the resulting shock wave, demonstrating its ability to travel through sand and rock. He estimated that the speed of the shock wave was about 500 m. a second but he was out by a large margin. Many years later, in 1876, Henry Larcom Abbot obtained a more realistic estimate for the speed of seismic waves in granite of 6.24 km a second.

Robert Mallet (1810–81) has been called the father of seismology (by no means the only one) and is one of the unjustly forgotten technological giants of the Victorian period and a distant relative of mine. He was born in Dublin, on 3 June 1810, the son of the owner of an iron foundry. After graduating in mathematics he worked with his father to build their business into one of Ireland's most important engineering companies, supplying ironwork for the expanding railway network, for the first Fastnet Rock lighthouse and a swing bridge over the River Shannon.

He also developed an interest in earthquakes and in 1846 presented a paper to the Royal Irish Academy. 'On the Dynamics of Earthquakes' is now considered to be one of the foundations of modern seismology – indeed, Mallet is credited with coining the words 'seismology' as well as 'epicentre'. He was especially interested in determining the energy unleashed by earthquakes, hence the gunpowder experiments, and with his son John undertook a series of experiments on how sound or energy moves through sand and rock, hence the explosion on Killiney Beach.

On 16 December 1857, Padula in Italy was devastated by the Great Neapolitan Earthquake which caused 11,000 deaths. It was the third largest earthquake known. Mallet was determined to study it first hand and petitioned the Royal Society of London and obtained a grant of £150 to go to Padula. The resulting report was presented to the Royal Society as the 'Report on the Great Neapolitan Earthquake of 1857'. It was a major work and made great use of the then new technique of photography to record the devastation caused by the earthquake. In 1862, he published the 'Great Neapolitan Earthquake of 1857: The First Principles of Observational Seismology' and suggested that the depth below the Earth's surface, from where the earthquake originated, was about eight or nine geographical miles. He concluded that an earthquake was due to a 'sudden flexture and constraint of the elastic materials forming a portion of the Earth's crust, or by their giving way and becoming fractured'.

He wrote:

When the observer first enters one of those earthquake-shaken towns, he finds himself in the midst of utter confusion. The eye is bewildered by 'a city become a heap.' He wanders over dislocated stone and mortar. Houses seem to have been precipitated to the ground in every direction of azimuth. There seems no governing law, nor any indication of a prevailing direction of overturning force. It is only by first gaining some commanding point, whence a general view over the whole field of ruin can be had, and observing its places of greatest and least destruction, and then by patient examination, compass in hand, of many details of over-throw, house by house and street by street, analysing each detail and comparing the results . . . at length we perceive, once for all, that this apparent confusion is but superficial.

He also scoured the libraries of Europe for information about earthquakes, and produced a list that was far more comprehensive than anything before it. He drew up a seismic map of the world and noticed that earthquakes were not randomly distributed across the Earth's surface but clustered in regions and in lines across the map. He had no explanation for it.

When I write about scientists I like to visit places of significance to them whenever possible, so one spring morning I travelled to West Norwood Cemetery, one of the 'Magnificent Seven' large cemeteries of London. It is a mixture of monuments, lawns, catacombs and a columbarium. In square 109 is a large Celtic cross. Robert is buried in grave no. 11023. In England no major earthquake will disturb him, but on his tombstone there should be a mention of earthquakes.

7

Tenham

One night in February 1869, Mr M. Hammond, owner
of Tenham Station, was camped with his brothers while
mustering cattle near the junction of Cooper and Kyabra
creeks. It had been an uneventful day and after their
evening meal they sat under the wide sky of south-west
Queensland – a land of cattle and cotton – drinking and
talking into the night. Suddenly, at about 2 a.m., there was
a flash and a streak of light across the sky, and 'a noise
like a rushing motor-car' was heard. Upon looking up the
brothers witnessed a brilliant meteoric shower passing
from west to east. Soon afterwards meteorites were found,
the largest weighing 130 lb.

A news article describes an eyewitness account of the
Tenham fall on Monday, 25 April 1880.

A few minutes after six o'clock [. . .] a very large and
brilliant meteor shot from overhead and descended in a
southerly direction. The meteor appeared to be the size
of a six-quart billycan, and was one splendid ball of fire;
it left no streak of light after it, and was the largest one
I have ever seen. When the meteor had descended about

three parts the distance from where I first saw it to the earth, I lost sight of it, as it was passing behind a large dark cloud. I stood looking in the direction the meteor was traveling, when a loud explosion took place in the same direction which slightly shook the ground for miles around; then a loud rushing noise could be heard as though a great blast of air was rushing through a large tube suspended in mid-air. This sound must have lasted for nearly two minutes when it died away. Next morning, when I rode up to Jundah, everyone there wanted to know what the explosion was, and the only conclusion we could arrive at was that when the meteor struck the ground it must have exploded, but we have not been able to account for the rushing sound afterwards. Inspector Sharp of the black troopers [an Aboriginal police force] said that when the explosion took place, the house he was in shook very much, and that when he ran out to see what was taking place he saw all the troopers running into the barracks with fright depicted on each countenance. From what I could learn I was the only white man at Jundah who saw the transit of the meteor. It is my opinion that it struck the earth a few miles above Galway Downs, and close to the Barcoo River, or we could not have felt the earth shake when it exploded.

Some years later it was reported that the Aboriginal people were 'deadly afraid' of the Tenham meteorites: 'They cover them in the bush with kangaroo grass, a twisted gidga bark and mud, and then by boughs over the top. Their idea is if the sun sees them, more stones will be shaken down to kill them', suggesting that they had witnessed the fall. There is scant information about what happened to the Tenham meteorites in their early days but one source says that they were taken by an opal dealer named T. C. Wollaston and sold to the British Museum in London, using a bogus story to explain their origins and how he acquired them. And there you can see them. The Tenham meteorite is on display in the Minerals Gallery on the first floor of the Green Zone, just above the shop in the Central Hall. How could those who had witnessed this dramatic event have known the secret inside the Tenham meteorites? This was a secret about the interior of the Earth that would remain hidden for a hundred years.

8

The Butler and the Housekeeper

'Everything is a matter of chronology'

Marcel Proust, *Remembrance of Things Past*

The development of seismographs continued. They were slowly getting better and approaching something that could become useful. In 1856, Luigi Palmieri installed his 'sismografo electro-magnetico' in the volcanic observatory on Mt Vesuvius. This instrument consisted of a conical mass on a spiral spring and was intended to give the direction, intensity and duration of an earthquake, and was capable of responding to both their horizontal and vertical motions. The mass was suspended over a basin of mercury. When a slight motion caused the tip of the cone to touch the mercury, an electric circuit was completed, which caused a clock to stop, indicating the time of the shock. In addition, U-tubes filled with mercury were used to detect horizontal motion. The closing of the electric circuit also started a paper recording surface and caused a pencil to be pressed against the surface. The recorder continued running until the paper was used up.

Palmieri's 'sismografo' seems to have been a crude but good earthquake detector for its time. It was used for many years on Mt Vesuvius and detected numerous shocks, but shocks further away were a problem. It was unable to detect many shocks that were felt in the nearby city of Naples. Interestingly, Palmieri observed that before Vesuvius was going to erupt, the 'shocks are more frequent; or to express it better, the ground trembles in a continuous manner with diverse phases'.

In 1889 came the breakthrough. The first record of a distant earthquake was made by Ernst Ludwig August von Rebeur-Paschwitz (1861–95), a remarkable seismologist who would no doubt have achieved even greater things had not his short life been plagued by tuberculosis. His seismic recording was made in the Prussian city of Potsdam, of an earthquake as far away as Japan. It stunned the scientific world. The instrument was a photographically recording horizontal pendulum. The 18 April 1889 publication of the journal *Nature* contained the following passage:

Dr von Rebeur-Paschwitz's letter, which appeared in *Nature*, vol. xl. p. 294, is of special interest to us in Japan, countenancing as it does the conjecture that the very peculiar earthquake felt and registered here on April 18 was the result of a disturbance of unusual magnitude. It was my good fortune on the day in question to be engaged in conversation with Prof. Sekiya in the Seismological Laboratory at the very instant the earthquake occurred. We at once rushed to the room where the self-recording instruments lay, and there, for the first time in our experience, had the delight of viewing the pointers mark their sinuous curves on the revolving plates and cylinders. At first sight it seemed as if the pointers had gone mad, tracing out sinuosities of amplitudes five or six times greater than the greatest that had ever before been recorded in Tokio. There was not much sensation of an earthquake; indeed, after the first slight tremor that attracted our attention, we felt nothing at all, although in the irregular oscillations of the seismograph pointers we had evidence enough that an earthquake was passing. Very few in Tokio were aware that there had been an earthquake till they read the report of it in the next day's papers. Thus the motion, though large, was too slow to cause any of the usual sensations that accompany earthquakes, and suggested a distant origin and a large disturbance, with a consequent wide extension of seismic effect. Excepting the slight tremors recorded at Potsdam and Wilhelmshaven, there has been, so far, no evidence of any such far-reaching action.

The event marked the true birth of the science of seismology. Only a few years later a network of forty seismic stations was set up in what was the beginning of global earthquake monitoring. This ushered in a new kind of scientist – the earthquake collector.

Perhaps the best of them lived not far from me. When I went searching for signs of his house, long demolished, on the Isle of Wight, I noticed some neglected cherry trees which puzzled me for a while. They are not native to the Isle of Wight so they must have been cultivated. Today there is almost no sign of Shide Hill House that stood just outside Newport less than a century ago. Houses are densely packed over the site though some of the outbuildings of the former estate can still be seen. The gatehouse still remains, looking forlorn and isolated at the end of St George's Lane. It's hard to imagine now but this was once the centre of the worldwide study of earthquakes. Cables and letters arrived daily giving details of major and minor tremors, near and far. Leading scientists, dignitaries and celebrities came to see the earthquake centre of the world and to take tea with 'Earthquake' Milne and his exotic Japanese wife, hence the cherry trees in what was once the garden. In a field nearby is a small memorial; among the inscriptions can be found the words, 'John Milne the Father of Seismology'. Seismology has become the major way we find out about the interior of the Earth. It grew out of the study of earthquakes, using devices, as we have seen, based on pendulums that swung when the Earth shook, giving a visible trace of the ground's vibrations. It was by interpreting these vibrations that scientists were able to determine the basic structure of the Earth, but before the observations could be analysed they had to be collected.

Professor John Milne, who lived and worked for many years at Shide Hill House, was part inventor of the horizontal pendulum seismograph and spent twenty years in Tokyo establishing the world's first earthquake laboratory, honoured by the Japanese but virtually forgotten by his own country. Soon after the Meiji Restoration in Japan in 1868 the new government began a programme for building lighthouses to enable trade to be restored with the west. They employed the Scottish engineer Richard Henry Brunton for the job and he soon realised that the lighthouses would have to be designed to withstand earthquakes. He urged the Japanese government to recruit British scientists to bring the latest ideas on the causes of earthquakes. Eventually John Milne, James Ewing and Thomas Gray developed prototype instruments that evolved into the modern seismograph. Seldom read is Milne's account of his overland trip to take a job at the Imperial College of Engineering in Tokyo. By train, by Mongolian camel route, on foot, his trek reads like that of Indiana Jones without the Hollywood gloss.

Milne began his studies in 1879, in the hope that earthquakes in Japan would be preceded by detectable noises, as the straining rocks responsible for the earthquakes were preparing to break. Some years earlier the Italian seismologist Michele Stefano de Rossi had suggested that waves were generated at the time of an earthquake by the lips of a 'volcanic fracture' moving rapidly up and down. A year later the Seismological Society of Japan was founded at Milne's urging following a strong earthquake in Yokohama. The society was the first devoted to seismology and its founding marked the beginning of a period of rapid growth of seismology in Japan. The Mino-Owari earthquake of 28 October 1891, the largest known to have struck the Japanese mainland, with its spectacular faulting, convinced Milne that faulting – the slippage of rocks against each other – caused earthquakes by the release of strain energy which had been stored in rock through the slow, pent-up deformation of the Earth's crust.

Using his new seismograph, at last scientists could look at the way seismic waves from earthquakes behaved and moved through the Earth. The shape of the waves – related to the arrival of various types of shock waves – held clues to the internal structure of the Earth. In a report on four small earthquakes recorded in the same month, one of Milne's co-workers detailed the anatomy of a seismic wave: 'The very gradual beginning and ending of the disturbance. In none of the observations did the maximum motion occur until after several complete oscillations had taken place. The irregularity of the motion. The successive undulations are widely different both in extent and in periodic time. The large number of undulations in a single earthquake, and the continuous character of the shock. The extreme minuteness of the motion at the Earth's surface.' For the first time, scientists had a description of earthquake motion, and it revealed a different manner of shaking than had been previously thought. Robert Mallet's widely accepted view that an earthquake consisted primarily of a single pulse of energy was shown to be incorrect.

Milne's seismograph was a remarkable instrument. It recorded photographically. Instead of having light reflected on to photographic paper with a mirror fastened to the frame of the pendulum, Milne had light shine on to the paper through the intersection of two mutually perpendicular slits. One of the slits was fastened to the solid pier. The other slit was fastened to the pendulum, and moved with the pendulum, causing the spot of light to move on the paper, thus recording the passing vibrations.

Milne knew that more seismographic stations were needed to monitor earthquakes wherever they occurred and to study the seismic waves that radiated from them. So he pressed for the establishment of a worldwide network of seismographic stations using standardised instruments. He was not the first to do this. In 1895, during the Sixth International Geographical Congress being held in London, Rebeur-Paschwitz made the same suggestion. Milne supported it but Rebeur-Paschwitz died a few months later and at future meetings where an international seismological association was discussed Milne rejected a suggestion that its international offices be in Europe, presumably because they would rival his work at Shide House. Soon splits appeared and factions developed, a scientific prelude to the fragmentation of Europe after the Great War.

Unsurprisingly Milne's seismograph was selected by a committee of the British Association for the Advancement of Science to be the standard instrument. By 1900, similar Milne seismographs were established on all of the inhabited continents. Milne knew that it was important to have a station in the Antarctic region so, in 1902, a Milne instrument was operated near the shore of the Ross Sea, at 77 degrees south latitude, as part of the British national Antarctic expedition of 1901–4. It recorded over one hundred quakes in the period of months in which it was operated.

Initially sixteen stations were regularly sending records to Milne and he set about analysing them. Using data from the Milne seismographs, and published data from

German and Italian observatories, he plotted travel-time curves for earthquakes with known epicentres. The first curve gave only the transit time of the seismic waves that stay near the Earth's surface but he was improving and a year later he had worked out more details about the shape of the waves that rippled through the Earth.

Milne was more than a scientist. He was an explorer, a keen naturalist and geologist and mining engineer. He enjoyed golf, music, literature and photography. Besides textbooks and many papers on seismology he wrote a number of science-based fiction stories as well as a best-selling humorous travel book. A picture of Milne the golfer hangs in Newport Golf Club and a trophy bears his name. It is said that he had the ability to mix at all levels of society, as this description from a local shows: 'I remember it as if it were yesterday and it was all of sixty years ago. I still have in my mind the squat figure of the old gentleman standing up there on the golf course behind his home, with that broad-brimmed hat of his, and his slight stoop, pointing out the houses on the other side of the valley, and making us laugh at the jokes he made as he explained their movement. He always spoke with a quiet Lancastrian accent which fascinated us lads, as did his nicotine-stained, bushy moustache with a gap burned in it by numerous cigarettes.' Milne would joke that he could calculate the time that the carts stayed outside the Barley Mow while their owners were in the pub by looking at his seismograph. One day he was excited but puzzled to find a series of enormous swings on his seismometer that he could not explain. A week later and at the same time of the day they appeared again. Milne eventually deduced that the records were made when the butler and the house-keeper were off duty together.

Milne's traces of the shock waves from earthquakes recorded all over the world, multiple traces taken a different distances from the same quake, were a scientific treasure and the key to understanding what was within the Earth. In 1899, Richard Dixon Oldham presented a thorough study of seismic waves and made a breakthrough in our understanding of them.

Oldham (1858–1936) is one of those relatively unknown scientists who deserve much wider recognition. His father was professor of geology at Trinity College Dublin so he was exposed to rocks from a very early age. He was educated at Rugby and the Royal School of Mines and in 1879 joined the staff of the Geological Survey of India. He is perhaps best known for his study of the Great Assam Earthquake of 1897 that for many years was the most comprehensive description of an earthquake available. But Oldham went further. As well as studying the earthquake itself he studied how other stations around the world detected it, looking at the seismic traces they had recorded and comparing those detected at different distances from the epicentre of the quake. Looking at those picked up between 4,300 and 4,900 miles he found three distinct wiggles in the data. He recognised three types of wave. The first two he suggested resulted from compressional waves (P waves) and, slightly later, shaking side-to-side waves (S waves), which are now called body waves; the third he identified as coming from surface undulations that had travelled right around the surface of the world.

But his interpretation, which we now know to be correct, was not able to be proven conclusively given the quality of the data available at the time. A decade later, Oldham analysed and compiled more observations from the Milne collection and in 1906 he published a paper that revolutionised our understanding of the Earth.

It was well known that the Earth had to be layered in some fashion. Isaac Newton two centuries earlier deduced the average density of the Earth, which turned out to be about twice the density of its surface rocks. Clearly things were denser and different down there. Oldham claimed to have observed and modelled the interior of the Earth using the paths P and S waves would take through the Earth, knowing that S waves – shaking waves – cannot pass through liquid and travelled at different speeds. Oldham saw that for a given earthquake parts of the world never received S waves from it. His explanation was that the Earth had a core of liquid iron through which S waves could not travel. For every earthquake there was an S-wave shadow caused by the Earth's liquid heart. Not everyone was convinced. Later, geophysicist and seismologist Sir Harold Jeffreys (1891–1989) proved that the Earth's core must have the behaviour of a fluid, and that therefore no S or shear waves can travel through it.

The oldest working seismograph, over a hundred years old but still fully operational, can be found at the University of Göttingen. It is the work of Emil Wiechert

(1861–1928) who was the world's second professor of geophysics. The first was Mauryey Pius Rudzki at the Jagiellonian University in Kraków. He suggested in 1898 that the Earth had a solid inner core. He carried out research in a laboratory built on the hillside of Göttingen's Hainberg where he started building seismographs. These days being a professor suggests an individual with a retinue of other academics, assistants and support staff. But things were different in Wiechert's time; in his department there was him, one assistant and a housekeeper! Today you can visit his 'Old Earthquake Vault' with its floor of concrete resting on a bedrock of shell limestone and see his 17-tonne pendulum and vertical seismograph in action. It still marks out its seismic traces on instrument paper covered in soot. In a paper published in 1910 Wiechert said that the further away from the earthquake the seismic waves were detected the deeper they had penetrated the Earth.

At Göttingen in 1908 one of Wiechert's students, Ludger Mintrop, destined to be one of the founders of modern geophysics, was the first person who worked out how to make sizable artificial earthquakes and went on to make a fortune with it, along with the development of portable seismometers. He used a 4-tonne steel ball lifted 14 m. into the air and dropped. Later he replaced the heavy ball with dynamite and with an array of portable seismometers created a three-dimensional picture of rocks a little way down; needless to say it proved invaluable and lucrative in the search for oil and coal. If you go to the University of Göttingen you can see Wiechert's impressive seismometer in action and on the first Sunday of every month, weather permitting, the Mintrop ball is raised and dropped.

Staying with the crust, it was clear that the seismograms held lots of information about the Earth just beneath our feet. Andrija Mohorovičić (1857–1936), the son of a blacksmith, who became a professor at the age of fifty-two, witnessed an earthquake near Zagreb. When he looked at the records from several seismometers set up in the region he realised that some of the shock waves from the earthquake were being reflected back to the surface from a boundary region between the crust and the mantle. It was announced in 1909 that it was between 5 and 10 km below the ocean floor and between 20 and 90 km beneath the continents. It is now called the Mohorovičić discontinuity, or the Moho, and scientists have long wanted to drill to the Moho and reach the Earth's mantle.

During most of the early history of seismology much interest centred on determining the depth of earthquakes. Some scientists said that the evidence of faults in rocks at the surface, and that some earthquakes had relatively small areas of damage, led to the conclusion that earthquakes must be quite shallow. But there were some suggestions that a few could be deeper. In 1913 G. E. Pilgrim looked at earthquake depths for ten quakes, including the 1906 San Francisco earthquake, which he found to have occurred at a depth of 140 km.

In 1922 Herbert Hall Turner (1861–1930), an Oxford astronomer and regular visitor to 'Earthquake' Milne on the Isle of Wight, who became a seismologist later in life, discovered a new type of earthquake that occurred much deeper than most other earthquakes, although many argued that the data was not altogether satisfactory. Incidentally, Turner is generally credited with coining the term parsec – a measure of astronomical distance – and it was he who forwarded to its Lowell Observatory discoverers the suggestion of an eleven-year-old Oxford resident that what was then considered the ninth planet be named Pluto. In 1928 the Japanese seismologist Kiyoo Wadati (1902–95) showed that deep earthquakes displayed a split distribution with peaks centred at about 400 km and 540 km, which we now know is because of changes to rocks as they descend into the mantle. He discovered what is today known as the Wadati–Benioff Zone, a region of intermediate and deep earthquake zones along oceanic trenches, which became the foundation for the plate tectonics hypothesis. Kiyoo Wadati is one of the significant but unappreciated seismologists because it would be fair to say that his work stimulated a revolution in thought about the mantle.

Some were still not convinced about the existence of deep earthquakes, pointing out that the typical timing errors in analysing the arrival of seismic waves were about a minute or two and the uncertainties in depth consequently large. Better evidence for deep earthquakes was obtained by Robert Stoneley and Fred Scrase in 1931. Some earthquakes, it seemed, came from more than ten times deeper in the Earth than others; there was no explanation.

In the following years more subdivisions were recognised in the crust but none as significant as the Moho. In 1924 Beno Gutenberg (1889–1960) looked at seismic waves travelling beneath the continents and the ocean crust and found significant differences, especially in the much thinner crust of the oceans. In 1929, in his book *The Earth*, Sir Harold Jeffreys discussed in detail the structure of the crust as it was then known. He concluded there were three layers and suggested that under the continents they were composed of granite, glassy basalt and a rock called dunite. As for what was under the oceans he concluded that they were 'less thoroughly studied'. Gutenberg, however, thought that Jeffreys' idea was too simple and that the crust was more complicated, pointing to some seismic waves that might be reflections from a layer above the Moho.

Despite probably being Europe's leading seismologist, Beno Gutenberg found it impossible to make a living in his science, having to run his father's soap factory during the week and carry out his research at the weekend. He was the obvious successor to Emil Wiechert but didn't get his job, probably because he was Jewish and anti-Semitism was on the rise in Germany. Like Albert Einstein, he fled Germany in 1933, taking a prestigious job at the University of California. It was there that he felt his first earthquake. The story has it that he and Einstein were walking across the Caltech campus, so deeply engrossed in conversation that they did not notice the ground shaking during the disastrous Long Beach earthquake of 1933. Gutenberg inherited his mother's musical talents. In California, Einstein often played the violin in chamber music at Gutenberg's home. In fact, Gutenberg often received royalties from publications in Germany in the form of piano scores.

9

Death on the Ice

As the twentieth century unfolded the Earth's crust began to give up some of its secrets. In May 1931 a team of German explorers were searching the Greenland Ice Cap in appalling conditions. They were looking for bodies. Six months earlier – before the start of the long polar night – their expedition leader, and brother of the lead searcher, had failed to return from the camp they called *Eismitte*, or 'mid-ice', situated at the very heart of Greenland. They hoped he and his companions had found some way to overwinter there even though they knew that the supplies would not have lasted that long. After 118 miles they found a pair of skis sticking upright in the snow and little else: no bodies. They struggled on to *Eismitte* to find one survivor. He told them that their leader and a companion had set off for base camp six months previously.

On 21 May they found his body. It was beneath the skies. He had been buried with care. He was fully dressed, lying on a reindeer skin. His eyes were open and they said later that he had a calm expression on his face. They believed he had died of a heart attack brought on by exhaustion. His companion, a Greenlander, had laid him to rest with respect. He must have set off alone and was never seen again; presumably he collapsed and lay unburied some distance away on the ice.

They left the body undisturbed and built an ice block mausoleum over it. Later they erected a large iron cross to mark the site. Before he left, Kurt Wegener said goodbye to his brother Alfred and vowed to continue the scientific observations. The German government wanted to bring the body back for a state funeral, but Else Wegener, Alfred's wife, knowing of her husband's love for the Arctic, refused and so he was left where his companion had buried him. It is gone now, having become part of the glacier, and probably lies 100 m. below the ice.

Alfred Wegener died aged only fifty but he left behind a grand idea, one of those ideas that change the face of science, one of those ideas that seemed so wrong to so many at the time but now appears so obvious that we wonder why anyone could have doubted it. He never saw it accepted but when it eventually was, twenty-five years after his death, it changed the way we view our planet and its interior. At some time in the distant future his body will be discharged into the sea following the glacier's long, slow and inevitable journey. Wegener's great idea involved the slow movement not of ice but of continents.

In 1911 Wegener came upon the idea that the continents moved around the globe; after all, the outline of Africa does appear to fit the outline of South America. The first person to notice this was Abraham Ortelius (1527–98), who was among the first generation to have a map of the world; indeed, he is generally recognised as the creator of the first modern atlas, the *Theatrum Orbis Terrarum*. Ortelius' thoughts were later described as suggesting 'that the Americas were "torn away from Europe and Africa . . . by earthquakes and floods"' and went on to say: 'The vestiges of the rupture reveal themselves, if someone brings forward a map of the world and considers carefully the coasts of the three [continents].'

Contrary to popular belief Wegener did not fit the shorelines of the continents either side of the Atlantic but used a contour that outlined their continental shelves, and obtained an even better fit. In 1912 he presented his thoughts on the idea of continental drift. He suggested that perhaps centrifugal force, or the wobble of the Earth's axis of rotation, could provide the driving force to move the continents. These suggestions could be easily shown to be inadequate for moving the continents. Wegener even suggested what we now know to be part of the correct explanation, that there were regions of sea floor spreading, writing, 'the Mid-Atlantic Ridge zone in which the floor of the Atlantic, as it keeps spreading, is continually tearing open and making space for fresh, relatively fluid and hot sima, rising from depth'. Unfortunately he did not work on this idea.

Wegener wrote a very good book presenting the case for continental drift but because the mechanisms he proposed for moving the continents were inadequate few people took the idea seriously; it was more than twenty years after his death before new data showed he was right.

The first evidence that showed Wegener was correct was obtained by the British geologist Keith Runcorn, the

Australian Warren Carey and the Canadian Ted Irving in 1956. New techniques had been developed to measure magnetism in rocks and how it related to the Earth's magnetic field. Rocks of different ages showed a different magnetic field direction. Some believed it was due to the poles of the Earth moving, but others thought that unlikely. The alternate explanation was that, instead of the poles wandering, the continents did. A few years later data from the mid-ocean spreading ridges clinched it for the continental drift idea; they showed symmetrical magnetic stripes either side of the ridge that could be attributed to changes in the Earth's magnetic field when the now separate rocks travelling away from each other had been formed in the mid-ocean ridge.

10

Superdeep

On most days for over a decade chief geologist Ivanovic
Vladimir Khmelinsky, then in his seventies, would drive
across the tundra in northern Russia on a forlorn journey
to one of the world's most remarkable scientific sites. He
was almost always the only person there. He would walk
through the abandoned building observing its inevitable
crumbling after each Arctic winter. It was beginning to
look ramshackle. There was a great deal of equipment
lying around rusting and there was a lot of core on the
surface that hadn't been properly logged. Sometimes he
would stop by the wellhead and recall past glories. He re-
membered 27 December 1983 as though it were yesterday.
The drilling had reached a record 12,000 m. They thought
they could go even deeper. They were wrong. Eventually
Khmelinsky could no longer make his pilgrimage. Three
years ago the site was demolished and the equipment
taken away. No money was available for maintenance.
The core material was removed to a research institute
near Moscow. All that remains is a pile of rubble and in
amongst it a welded plate on the ground.

There is no monument, no acknowledgement that this

was one of the great scientific projects of the twentieth century, involving twenty-five years of work, with over 400 people, including drillers, technicians, geologists and support staff. Nothing to mark what had been once called the Soviet equivalent of a moonshot. Ivanovic Vladimir Khmelinsky was heartbroken, as were many others. It was called the Kola Superdeep project and its intention was to drill through the crust of the Earth into the mantle. They were not the first to try, or fail.

In the late 1950s and early 1960s the United States National Science Foundation – the body that chooses and funds major scientific projects – provided some money to Project Mohole, which was to drill through the relatively thin oceanic crust to reach the Mohorovičić discontinuity. The idea came chiefly from one of the pioneers of plate tectonics, Harry Hess, and Walter Munk, who wanted to get stuck into a really big landmark project. At breakfast one morning in 1957 at Munk's La Jolla, California, home the idea of Project Mohole was proposed. It was an ambitious project, for no oil company had considered deep-water drilling or even really contemplated the technical problems of keeping a ship steady while drilling into the ocean floor. In 1961 five exploratory holes were drilled off the coast of Guadalupe Island, 250 km off the west coast of Mexico, on the theory that the crust was thinner beneath the ocean. They lowered the drill bit through 3,800 m. of water, 170 m. of sediment and returned a few metres of basalt. It was reported in *Life* magazine (14 April 1961) by the novelist and amateur oceanographer John Steinbeck, who was aboard the drilling ship. In the

91

face of escalating costs and poor management, however, the project was abandoned after a few years.

A few years later Soviet scientists, seeing that this scientific goal still remained, and perhaps smarting over the Americans landing on the Moon, proposed a series of superdeep holes to explore the nature of the Earth's crust under the USSR. The project's director was Oktan Ibragimov, who announced that five holes would be drilled across all of the USSR, in the Ukraine, the Arctic, the Urals, the Kurile Islands in the Pacific and at Saatly, near the Caspian Sea, and it would go down to 15 km, the estimated depth of the mantle in this region. Ibragimov thought that they might even reach the mantle in 7 km. Preliminary work had also been done at the Kola Superdeep site, on the Kola Peninsula. Drilling began at Kola on 24 May 1970.

To get to the site of the world's deepest hole requires a journey through what may well be the most depressing, most polluted landscape on Earth. It is high in Arctic Russia near the border with Norway. On the Norwegian side of the border the Arctic is pristine and wonderful, especially in the spring. Wooden houses, painted in bright colours, are scattered around the outlets to the sea. The roads are very good and everywhere there is a quiet sense of Nordic order. Cross the border into Russia and it's a quite different matter.

When the area was first surveyed by geologists in the 1920s they found evidence of nickel-bearing rocks that became essential for industrialisation under Stalin. Within a few years workers were forcibly shipped to the area to labour in dirty and unsafe mines not far inland from the important naval base at Murmansk. In 1946, following the retaking of the area by the Soviet army, the number of open-cast mines and smelting shops increased. Then and now the toxic chemical by-products of smelting pollute the air, causing health problems and blighting the land-scape. Grim apartment blocks cluster around meagrely supplied shops. Haggard workers were hemmed in by military checkpoints. Westerners were not allowed access.

Nowadays it is possible to drive from Norway to the region without a passport. The military checkpoints have gone, for this is now part of the E105 trans-Europe motor-way route. But the pollution is not so easily removed. You drive past the military settlements at Pechenga to settle-ments in the north-west. They are divided into two sites, one at Zapolyarny and one at the settlement called Nikel. The nickel-smelting plants released almost a hundred thousand tons of sulphur dioxide in 2008, five times more than the entire Norwegian emissions. Nikel is probably the most polluted town on Earth. The sulphur dioxide from the smelting process doesn't come out of chimneys but, rather, seems to seep through the walls. The place smells of burning rubber and sulphur.

Finally, you travel a few kilometres over a barren tundra of low shrub down a road that is disintegrating. This attempt to drill through the Earth's outermost layer man-aged just over 12 km before it was abandoned in 1992 as the temperature at the drill tip touched 200 degrees C, resulting in insurmountable technical problems. Scien-tists believe it only reached a third of the way through the continental crust. The deepest borehole, called SG3, drilled through the sedimentary-volcanic sequence of the lower Proterozoic Perchenga formation, ending at 6,842 m., and then through part of the Archean granitic-metamorphic complex.

One of the most remarkable discoveries made by the borehole was the absence of the transition from granite to basalt at a depth of between 3 and 6 km. Seismic wave observations suggested a discontinuity existed at this depth, probably due to a change in rock type, but there was no such transition, only more granite. The explanation was that the discontinuity indicated by the seismic waves was due to a metamorphic change in the rock, rather than a change in rock type. More surprising was the fact that the rock had been thoroughly fractured and saturated with water. Free water was not supposed to exist at such depths. It could be that the water consisted of hydrogen and oxygen that was squeezed out of the surrounding rock and retained due to a layer of impermeable rock. Drillers also described the mud that flowed out of the hole as 'boiling' with hydrogen: such large quantities of hydrogen were highly unexpected. But perhaps the most remarkable thing the superdeep borehole found was microscopic plankton fossils in rocks over two billion years old, 6 km beneath the surface. There were twenty-four ancient species in total encased in organic compounds which somehow survived the extreme pressures and temperatures.

Cornelius Gillen, now at the University of Edinburgh, was on an international panel of scientists chosen to evaluate

and guide the Kola Superdeep project, which he looks
back on with sadness. 'It ground to a halt due to lack of
government support and then the stream of publications
from the project seemed to get stuck. It was a difficult time
for all the workers, as for many it was their life's work.'
The head of the project, David Guberman, hoped that
it would become an international research facility, but
found he couldn't pay staff salaries. He died shortly after
its closure. Some workers will still tell you it was a cursed
operation and say that when the hole was really deep
some of those at the wellhead talk of screaming coming
from below as if they had pierced the roof of hell. It's hard
to say if they really believe this, or if it's just for the benefit
of visitors. The Kola experiment failed to reach the mantle
of the Earth. But there are plans to try again.

In the past few years scientists have once again thought
about drilling to the mantle, thanks to better technology
and a better understanding of the rocks beneath our feet.
Surveys have been carried out at several Pacific sites to find
one that might be used for the deep drilling. But whatever
the chosen site it will require a great deal of money, time
and equipment. Getting through the oceanic crust to the
mantle would provide geologists with a great deal of valu-
able information about our planet's outer skin.

In 2002 the Japanese launched the drilling ship *Chikyu*,
a giant of a vessel capable of carrying 10 km of drilling
pipes and of drilling in 2,500 m. of water. New drill bits
and lubricants will be needed to withstand high pressures
and temperatures up to 300 degrees C. To find the best
site for drilling a number of factors must be considered. It

would need to take place in the shallowest possible water overlying oceanic crust, which means going as close as possible to the mid-ocean ridge where new crust is formed. It should also be in the coldest possible crust, which is away from the ridge. This limits the possible sites to three – off the coasts of Hawaii, Baja California and Costa Rica. The site near Hawaii, for example, is the coolest, but also the deepest, and close to recent volcanic activity, so there are pros and cons for each site. The project might be taken up again, or it might not. As Harry Hess told a US National Academy of Science meeting in 1958, when defending the first Mohole project: 'Perhaps it is true that we won't find out as much about the Earth's interior from one hole as we hope. To those who raise that objection I say, if there is not a first hole, there cannot be a second or a tenth or a hundredth hole. We must make a beginning.' While the new Mohole project takes its chances there is a place where nature has already done it for us.

The Iapetus Sea was formed 600 million years ago when a giant continent started to break up. In many ways it can be seen as a precursor of the Atlantic Ocean. A new super-continent called Laurentia was being formed, and soon the floor of the Iapetus Sea was being driven beneath it – a process called subduction. But this time the ocean floor did not simply descend beneath the lighter rocks of Laurentia. The floor split and some of it rode over the continent, taking with it parts of the mantle. After hundreds of millions of years the overlying rocks were weathered away, meaning that the tablelands of western Newfoundland are one of the very few places on Earth where, thanks to a geological accident, you can walk on exposed mantle rocks.

The rocks are reddish brown because they have a high iron content and, having been exposed to the atmosphere, they've rusted. One guide who shows visitors the best of the outcrops and explains why they are special told me that being next to these rocks was a strange experience: 'you can hold them, even taste them if you like, but you can't take any away.'

You can still find the remains of the Iapetus Sea elsewhere on Earth. The mountains that bounded it are still around, though not as grand as they once were. The southern part of that range is now the Appalachian Mountains that straddle the United States while the northern section is found shared between Scotland and Scandinavia.

We are reaching the end of the first part of our journey and are soon to leave the crust behind. We are leaving the lithosphere as well, the rocky outer crust and part of the upper mantle that is relatively cool and brittle, and heading towards the asthenosphere, a hotter and more mobile region where rock flows and in which no earthquakes should occur because rock flows under strain. But some earthquakes do occur in the asthenosphere, and solving that particular mystery will guide us down deeper into the Earth.

11

The Hansbach

After taking a wrong turn they run out of water and Axel almost dies of thirst. Retracing their steps Hans hears water rushing behind a granite wall. He presses his ear against the warm rock trying to find the place where the rushing sounds the loudest. He strikes the rock with his pickaxe and out gushes hot, steaming water, which they let cool. They called the stream the Hansbach and they followed its course deeper towards the centre of the Earth.

We have a river to follow in our journey to the centre of the Earth, but our Hansbach is not a river of water but a river of rock, a slab of ocean floor, to be precise, that dives downwards. We have seen that the continents move and that the surface of the Earth is divided into tectonic plates and that sometimes the ocean floors descend below continents as tectonic plates collide. It is this process, subduction, that connects the deep Earth to the surface. There are twenty-four distinct subduction zones stretching 55,000 km across the face of our planet and each one is slightly different. Most of them define the 'Ring of Fire' as the volcano and earthquake zones that rim the Pacific but there are others. A particularly active small subduction zone is found near the South Sandwich Islands. Each year, as the ocean floor disappears, some 3 square km of surface are lost and replaced by new land formed at mid-ocean ridges. The mid-Atlantic mid-ocean ridge allows the South American and African plates to move apart, pushing South America into the Pacific plate at about 25 mm a year. The East Pacific plate is fighting back. The Nazca plate is sliding under South America, which – because of the forces acting on it – has become a very thick continent, up to 70 km in some places. The floor of the Indian Ocean is spreading, pushing India towards Asia at about 15 mm a year, the collision creating the Himalayas – the greatest mountain range on Earth. In an analysis of the forces acting on the subduction slabs it was found that they are pushed apart by the upwelling at a mid-ocean ridge when new land is formed but there is a much stronger pull produced when the slab subducts. This would explain why the plates that do not have subduction zones associated with them move more slowly than those that do, about the speed of nail growth as opposed to the speed of hair growing.

Professor Bob Stern of the University of Texas is a leading authority on subduction. He calls it the most important solid-earth process and he is right. He believes subduction zones are generally not appreciated and are immense in scale and importance. Raised in California, Stern's interest began when as a graduate student he wanted to visit the Mariana arc subduction zone in the late 1970s. He says it started his love affair with subduction zones. 'You can't really visit them,' he says, 'we can see what's happening on their roofs and the chimneys coming out of them, but finding out what is actually going on is more difficult.'

The thin and dense ocean floor of the Nazca plate is descending under South America at a region called the Peru–Chile Trench. It dips down at 30 degrees, leaving the surface behind.

Where is it going?

On their journey deep into the Earth, Professor Liden-brock and his companions encountered 'crystals . . . like globes of light'. Jules Verne couldn't have known how, 150 years later, the study of crystals would shine a new light on the deep Earth.

We are following our Hansbach – the ocean floor, a lithospheric slab – that has begun its descent into the Earth. Heavier than the surrounding rock, as it comes up against lighter rocks it dips beneath them and now nothing can stop it. It was the ocean floor as it moved away from its birth at a mid-ocean ridge, but now it begins a new life underground, possibly not returning to the surface again for hundreds of millions of years, or perhaps billions, or perhaps never.

The slab is heading through the crust into the mantle – the largest region of our planet, comprising about 82 per cent of its volume and 65 per cent of its mass. The mantle is where the archaeology of our planet is stored along with many mysteries, newly recognised ancient structures and processes that may be necessary for life to exist on the surface. The Earth's past is down there, and rising from its depths at a glacially slow speed is possibly the greatest environmental challenge mankind will face from our planet, leading to events that will dwarf any earthquake or volcanic eruption we have yet witnessed.

When I was young and majoring in physics I went to lectures about the structure of the Earth. Much detail was given about the crust and the core but the mantle was invariably passed over as a bland, featureless region, just a layer of rock between the crust and the core in which rock moved very slowly in a convection pattern transporting energy outwards to the surface, providing the energy to move the continents. It wasn't where the action was. There were a few layers within it, all at the top, and nothing else. It seemed that the greater part of our planet was somewhat boring. We should have known better! Many geophysicists passed it by, seduced by the sweeter physics found at the heart of the Earth. It is a feeling that Professor Ed Garnero of Arizona State University shared when he attended geophysics lectures.

He told me: 'For many years only the mantle's upper section was considered interesting. Increasing pressure and heat were supposed to have erased any structure below creating a homogeneous mass of slowly moving rock reaching downwards until the dramatic boundary with the outer core is reached. I believed all of this when I was a new researcher in the 1980s. I was taught that mantle structure, what there was of it, was radial, layers and layers of differing rock composition as the atoms in those rocks yielded under pressure to form ever closer arrangements.'

Now that we know differently I asked Professor Garnero what advice he would give his younger self knowing now just how dramatic the Earth's mantle has become. 'The evidence was there that the mantle was not so simple, but generally scientists ignored it and they passed that on in the lectures I attended. The new data that was eventually to change everything was beginning to emerge, but it didn't fit their preconceptions of what the mantle must be like. It is often the case when the subject changes. The signs were there in the seismic data, you could see there was something more. You could see the reflections from various layers were not clear-cut – there was what we thought was noise or scatter in the data as well. Importantly that scatter in the data was different in different parts of the world. This was explained by layers in the mantle being at slightly different depths.' But as Ed Garnero looked at the blurred seismograms he, and others, began to wonder. 'If I could go back and talk to myself when I was starting out,' he says, 'I would say look at that data carefully. It is not noise. It is real. It represents real reflections from real structures in the mantle. It is telling us that not only does the mantle have radial structure, but there is side-to-side structure as well. It is saying that the mantle is far more interesting and important than we all realised.'

The shifting tectonic plates that move across the face of our planet came early in its history, not long after the Moon, perhaps. Indeed, we have seen that in the zircons from the Jack Hills and in the ancient greenstone rocks on the shores of Hudson Bay there is evidence that the basic mechanism of plate tectonics – subduction – had already started to some degree. It may be that these interlocking slabs of crust that float on Earth's viscous upper mantle were made when they were pulled downwards, fracturing the surrounding crust. After it had happened many times the weak areas formed plate boundaries. Heat from the Earth's interior caused the mantle to stir slowly, transporting the energy to the surface where some of it was used to move the plates. Perhaps this explains why Venus – a world almost identical in size to the Earth – does not have tectonic plates. Conditions on Venus are much warmer, perhaps allowing the crust to heal after a piece sinks into the mantle.

So the new rocks formed at mid-ocean ridges spread sideways becoming slabs of moving basalt. On the sea floor they often gather sediments that can pile up in a thick layer. But its life on the Earth's surface is a relatively brief one, no more than 200 million years, for these rocks so recently solidified from the molten mantle will rejoin it, bringing cold from the surface, sediments and great changes. At subduction zones these slabs bend and enter the mantle at various angles, sometimes perhaps 10 degrees, sometimes much steeper. Under the Marianas it's going vertically. Using seismic data scientists can see these slabs as they descend and follow them several hundred kilometres into their dive. Take the Tonga subduction zone where the Pacific plate is descending beneath the Indo-Australian plate. It dips at an angle of 45 degrees and can be followed in the P-wave seismic data to a depth of about 700 km. As it descends, above it are two ranges of volcanoes with a wide plain between them called a spreading centre. When the region's earthquakes are mapped a curious pattern emerges because they follow the slab down.

All subduction zones, says Bob Stern, are individuals. They have a basic pattern and much variation. The simplest view is of one plate subducting beneath another but things are often not that straightforward. Take Japan, for instance, perhaps the world's most earthquake-prone country. Beneath central Japan the Philippine Sea plate and Pacific plate are both subducting beneath the Eurasian plate on which sits Tokyo. This is the cause of deep-thrust earthquakes, the most powerful and feared type of earthquake. The situation in Japan is, however, even more complicated, as recent analysis of seismic data has identified that a fragment of the Pacific plate has come away and is jammed between the Pacific and Philippine Sea plates and is overriding the Eurasian plate, and all this is happening directly under Tokyo. Indeed, the position of the broken slab coincides with the location of recurring deep-thrust earthquakes beneath the metropolis.

Indonesia has long fascinated those interested in tectonics because it is sandwiched at the junction of four important tectonic plates. Geophysicists are especially attracted to the curved island arcs around the eastern end of the Banda Sea. Stretching from Timor in the north to Buru in the south, there are two parallel arcs of islands bending through 180 degrees with a radius of about 200 km. The inner chain is volcanic, the outer is not. At the eastern end, between the inner and outer arcs, is one of the world's deepest basins – the 7-km-deep so-called Weber Deep – a depression on the sea floor between the subduction zone and what is termed the forearc region of the subduction region. As Bob Stern notes, we can see many things happening on the roof of such a zone. Seismic activity in the region is also curved around the islands. The seismic history of the region includes some of the largest earthquakes ever recorded at depths of around 100 km below the surface.

The Izu–Bonin–Mariana arc is a particularly outstanding example of subduction. It extends for almost 3,000 km southwards from Tokyo. It's the eastern part of the Philippine Sea plate and it boasts the deepest gash in the Earth's surface – the Mariana Trench – nearly 10,000 m. and where mankind has come closest to the centre of the earth. Here the western Pacific plate – a basalt sea floor created in the mid-Jurassic and early Cretaceous geological periods – is being subducted. There are arcs of volcanic islands close to the subduction zone. Looking closely at what is going on again reveals things are more complicated than at first sight. Split from the Philippine Sea plate is a smaller one called the Mariana plate. Basic pattern; many variations.

But why do some slabs plunge straight down and others take a shallower angle? Young oceanic slabs, less than about seventy million years, often subduct into the Earth at twice the rate of older ones. It's thought that older crust, being dense and strong, cannot bend very quickly and so enters the mantle at a shallower angle than younger, lighter and weaker slabs. This difference might affect their behaviour further down, as we shall see.

Most subduction zones appear to be young and from this scientists surmise that since subduction has been taking place for billions of years it must be fairly easy to make a new one as old ones die out. For example, the Izu–Bonin–Mariana system probably began many tens of millions of years ago as a widespread foundering of dense sea-floor rocks in the western Pacific which formed one of the largest subduction zones we see today.

The forces involved in a slab of oceanic crust overlaid with sediments, perhaps in total a hundred kilometres thick and several hundred kilometres wide, are titanic. As it descends below the continental crust, because it is denser than the rock beneath it, some of the sediment is scraped off, forming what is termed an accretionary prism – a rich fertile strip such as the California coastline that attracts people even though it is inevitably an earthquake zone. Sometimes there isn't very much in the way of sediments and the scraping of the subducting plate fractures the leading edge of the continent, forming rifts and mountain ranges such as the Andes.

One thing to note is that as the slab starts to descend it is cold and the mantle is hot. This has major implications for what will happen, according to Steven Hauck II of Case Western Reserve University: 'Because the temperature inside the slabs is cold, as they had been at the surface of the Earth for tens to hundreds of millions of years, as they start to descend into the mantle they begin to get warm, but it is still way colder than the surrounding mantle.' It is also compressed by great forces causing fundamental changes to the structure of the rocks, and it is brittle and fractures and fragments, causing earthquakes. Earthquakes always happen where rocks are strong enough to break. If they're weak they flow. But what of the volcanoes? Why are there arcs of volcanoes just beyond a subduction zone? Most of the earth's volcanoes are associated with subduction zones and they go hand in hand with earthquakes. A quick glance at the so-called Ring of Fire around the Pacific shows that subduction zones, earthquakes and volcanoes are co-sited.

Water and fluids containing water have a big influence in the Earth's crust and upper mantle. They control rock strength and hence the incidence of earthquakes. Water also moves chemical elements from place to place and forms ore deposits. As the slab descends and large quantities of rock are transported to increasing depths, the temperature and pressure increase to levels that the slab cannot withstand and changes occur. As its temperature increases minerals containing water such as epidote and chlorite break down and release fluid. Exactly where this happens depends upon how fast the rocks descend. The water seeps out of the slab and reaches the mantle rocks above it, where it causes the melting temperature of those rocks to lower, sometimes by as much as 400 degrees. This rock, now much less viscous, makes its way to the surface where new volcanoes are formed. That is why a hundred or so kilometres beyond a subduction zone there are arcs of new volcanoes that will be there as long as the descending slab gives off water. Such volcanoes are very different from the other kind found on the Earth's surface at mid-ocean ridges, in rift valleys and sometimes all on their own.

Some geophysicists speculate that we are entering a new phase of subduction because as time goes by and the Earth cools more water is taken into the Earth. We have seen that as the oceanic slab makes its way from mid-ocean ridge to subduction zone it becomes permeated with water that gets into deep cracks and into the very composition of the rock itself. While much of this water is released, resulting in volcanoes, it is thought that some water should remain trapped in the slab as it descends deeper into the mantle. A clue as to how much water is taken so deep comes from studies of the amount of water in the Earth's oceans over the past 500 million years or so. By looking at sedimentary rocks scientists are able to estimate how deep the ocean was when they were laid down, an estimate to within a factor of two or three. The result is that not much water has been lost underground, but that might not be the case in the future.

The Earth is cooling and the mantle is cooler than it was 500 million years ago. That means that not as much water is wrung out of the rocks as the slab descends, so in the future more water should be removed from the ocean and stored in the Earth. Some scientists point to the Tonga and the South American subduction zones as being cool enough for this to happen already. It's estimated that over the next 100 million years the global sea level could be reduced by about a hundred metres as water is moved into the Earth's interior. Slowly, over vast swathes of geological time, the oceans might decrease in volume and the face of our planet might change because of this subduction side effect.

The slab is getting cooked, especially the thin layer of sediments it carries down at its top. The sediments can get to 1,200–1,300 degrees C, producing some very interesting rocks. It reaches the deepest point of seismicity, about 670 km down, after about ten million years but the absence of earthquakes at deeper levels shows that the heat has reached its centre and the rocks are no longer brittle enough to make earthquakes. If you think that is the end of the slab and that it melts and merges with the rocks around it, like chocolate slabs melting into hot chocolate, you could not be more wrong. The slab's life is far from over. It will soon reach the first major structure in the mantle called the transition zone.

The transition zone was first glimpsed during the 1920s and 1930s, when great strides were being made in analysing seismic data even though there were not all that many seismic monitoring stations. It was noticed that there was a change in seismic speeds somewhere in the upper regions of the mantle. In 1936 it was suggested that it was the possible depth of the transition under pressure of the mineral olivine (magnesium or iron, silicon and oxygen) to a denser structure in which its constituent atoms were more closely packed and better able to withstand the pressure at those depths. The American geophysicist Francis Birch (1903–92) then classified mantle seismic structure in terms of composition, resulting in his landmark 1952 paper. Birch is perhaps best known for his part in the bombing of Hiroshima and Nagasaki and his participation in the Manhattan Project. He developed the gun design for the nuclear weapon and helped load it aboard the *Enola Gay*. He spent his entire academic career at Harvard. His doctoral adviser was Percy Bridgman. He was a pioneer of solid-earth physics and how the study of the physics of minerals and rock, combined with seismic data, could lead to an understanding of the composition of the Earth. He concluded that the upper mantle is made of the minerals olivine, pyroxine (calcium, sodium and iron) and garnet (composed of olivine and pyroxine) and that the lower mantle is composed of denser material such as periclase, a form of magnesium oxide. Between them he also thought there would be a transition zone where pressure altered the packing of atoms in the minerals. At the time it was a suggestion that was well ahead of seismic or mineral physics.

It was thus that seismic observations and mineral physics started to merge. The seismic data suggested changes in the structure of minerals at certain depths and hence certain pressures inside the Earth, and it was up to physicists to build equipment able to reproduce these pressures in the laboratory. It was not an easy task.

12

The Pressure Principle

When they study matter scientists have many options. They can measure dimensions and weights, cool and heat, stretch or squeeze, and they were not slow in trying to squeeze rocks and minerals. Victor Moritz Goldschmidt (1888–1947), born in Zurich to Jewish parents, and Vladimir Vernadsky (1863–1945) are considered the fathers of modern mineralogy. Goldschmidt was a professor in Norway during the German occupation. Norway was important to Germany as a U-boat base and for securing iron shipments from Sweden. Historians have described the German occupation of Norway as brutal. This applied particularly to its Jews: 736 of the original Jewish population of 1,800 in Norway were deported. Goldschmidt was arrested on 26 October 1942 and released on 5 November. He was arrested again on 25 November and was on the pier awaiting transport to Auschwitz when he was held back: perhaps his science would be useful to the Third Reich as he was working on the properties of plutonium. He later escaped to Sweden and then England, returning to Norway in 1946 but dying shortly afterwards. Today his name is given to the Goldschmidt

Conference organised each year by the European Association of Geochemistry.

But it was the Nobel Prize-winning Harvard physicist Percy Williams Bridgman (1882–1961) who ushered in the age of high pressure by subjecting to it almost anything he could get his hands on. Many years after Bridgman's death Harry Drickamer (1918–2002) said of him, 'Many people have asked me whether I was a student of Bridgman. Formally, of course, I was not. However, in a very real sense we all are students of Bridgman. We all share his enthusiasm for high-pressure research. We can all hope to reflect something of his intellectual honesty. Today we still use his techniques, or modifications and extensions of those he developed ... after his death, his spirit and his ideas still permeate the field. Seldom, if ever, has one man been so significant for any important area of research.' Bridgman even discovered a new kind of ice by compressing water. Drickamer also said, 'I'm sure Professor Bridgman would be most pleased to see the extent to which high pressure has become an integral and essential technique in modern physics, chemistry, geology, engineering, and biology.' And no one did more to spread this influence than Harry Drickamer.

In 1955 Robert Wentorf Jnr (1926–97) of the General Electric Laboratories in New York went out to buy some peanut butter (crunchy). He didn't use it in a sandwich but took it back to his lab, subjected it to intense pressure and turned peanut butter – major component carbon – into tiny crystals of diamond. Such is the power and the wonder of pressure. It is the force that helps power the stars and shapes the planets, and the pressure of the atmosphere on the surface of the Earth keeps us alive.

Welcome to the wonderful world of pressure. First, some reference points. The pressure at sea level – one atmosphere – is about a kilogram per square centimetre. Underneath the ocean the pressure increases by one atmosphere every 10 m., so at the bottom of the Mariana Trench – which, at almost 11,000 m., is the deepest part of the ocean – the pressure is 110 atmospheres, or about eight tons per square inch. Scientists measure pressure in pascals (Pa). Place a dollar bill on a table and it will exert a pressure of about one Pa. A strong breeze produces 100 Pa and a hard punch about 200,000 Pa. The carbon dioxide pressure in a champagne bottle is about 500,000 Pa. Your bite produces a pressure of a million, or 1MPa. Inside the Earth the standard measure is a gigapascal, or a thousand million, one GPa being the equivalent of 100,000 atmospheres. The centre of the Earth has a pressure of about 350 GPa, or thirty-five million atmospheres. It might sound extreme but it is we who are a rarity in the universe. Very little matter is at a pressure of one atmosphere; most is either in the form of very thin gas spread between the galaxies and the planets or at pressures greater than 100,000 atmospheres inside dead stars and planets. Percy Bridgman started with machines that could produce pressures of only 0.3 GPa and made machines capable of 10 GPa, or about 3 per cent of that at the Earth's core.

Percy Bridgman had made a start but there was a long way to go. The puzzle was how would minerals made of just the handful of elements that comprised the vast bulk of the mantle – silicon, oxygen, calcium, magnesium and iron – behave under such pressures and temperatures. Australian geophysicist Ted Ringwood (1930–93) took up the challenge. Early in his scientific career he spent several months at Birch's lab and became fascinated by the Earth's mantle and wanted to recreate the conditions there to see for himself what happened. He was able to predict that the pressure-induced changes in the mantle of the minerals olivine and pyroxene should occur in the mysterious transition zone. Working at the Australian National University, Ringwood built presses capable of squeezing olivine at 5–6 GPa. By the early 1960s he led the most active research group in the world, along with S. Akimoto of the Institute for Solid State Physics at the University of Tokyo. He generated high pressures using giant presses with opposed tungsten carbide pistons of various designs.

As Ringwood was doing this, techniques of analysing seismic waves had improved; by the 1960s seismologists were beginning to experiment using arrays of seismometers to provide more data and the fuzzy mantle transition zone was resolved into two regions, one at a depth of 410 km and another at 660 km. The 410 km shows a sharpness of 2–4 km beneath oceans and up to 35 km beneath continents. The 660 km discontinuity is much thinner, less than 5 km, and is the sharpest change seen in the mantle. These changes in seismic speeds implying changes in the density of the rocks were far too sharp to be due to compaction; something had to be happening to the very structure of the minerals themselves.

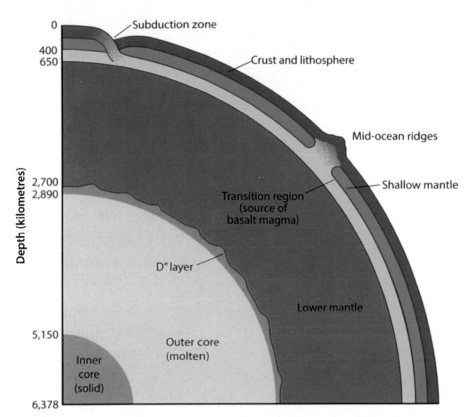

The internal regions of the Earth. A thin crust overlays the mantle – the Earth's most extensive region. Beneath it is an outer core of molten iron and a solid-iron inner core.

A 4.4-billion-year-old zircon. It is the oldest known piece of the Earth, a tiny fragment of its first crust.

The oldest rocks in the world exposed
on the shores of Hudson Bay, Canada.

◀ The Jack Hills in Western Australia, site of the world's oldest zircons. They have enabled scientists to determine the conditions on the surface of the Earth just after it formed.

▼ The deepest sample ever retrieved from a well, taken 12 km below the surface during the Kola Superdeep project.

The deepest hole ever drilled into the Earth – the Kola Superdeep Borehole in northern Russia that reached a depth of 12,262 metres in 1989.

Emil Wiechert (1861–1928) produced the first model of a layered Earth, and became the world's first professor of geophysics.

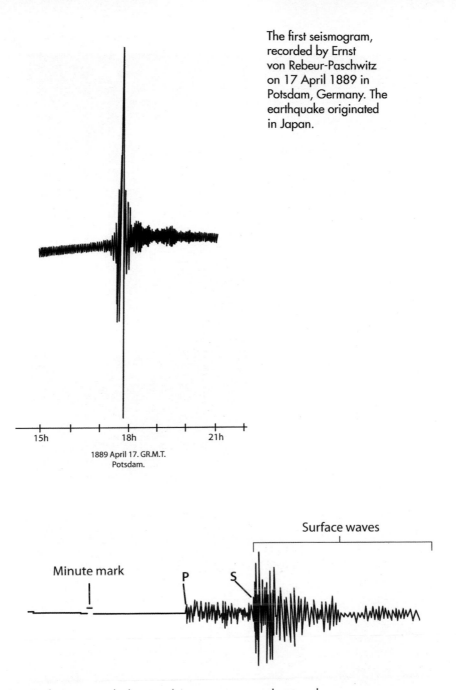

The first seismogram, recorded by Ernst von Rebeur-Paschwitz on 17 April 1889 in Potsdam, Germany. The earthquake originated in Japan.

15h   18h   21h

1889 April 17. GR.M.T.
Potsdam.

Surface waves

Minute mark   P   S

A typical seismograph showing the pressure waves that travel through the Earth after an earthquake.

John 'Earthquake' Milne (1850–1913) and
his Japanese wife Tone on the Isle of Wight.

Richard Dixon Oldham (1858–1936) made the first clear identification of P waves, S waves and surface waves in seismograms, and also the first clear evidence that the Earth had a core.

Inge Lehmann (1888–1993), discoverer of the Earth's inner core. In 1936, using earthquake data, she proposed that the Earth's core was not all molten iron but had a solid centre.

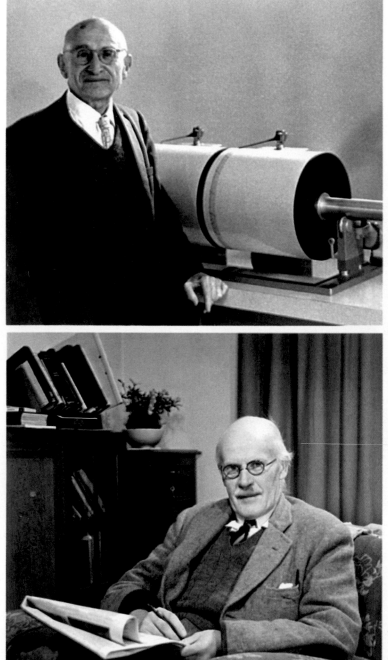

Beno Gutenberg (1889–1960) made many pioneering discoveries in seismology. Along with Charles Richter he developed a scale for measuring the intensity of earthquakes.

Although he opposed the theory of continental drift, Harold Jeffreys (1891–1989) made seminal contributions to the study of the Earth's interior.

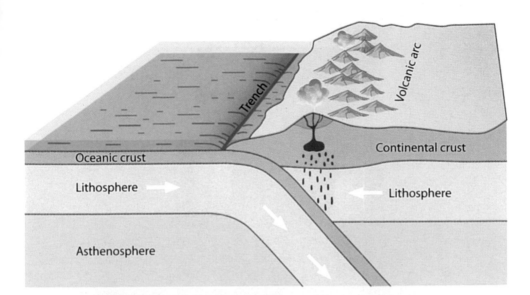

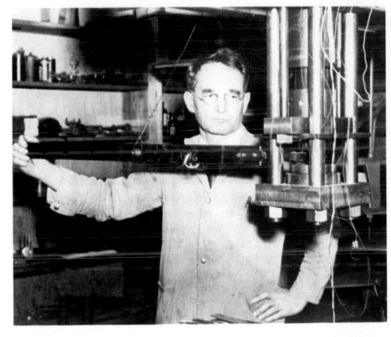

∧ The most important solid-earth process: subduction, whereby oceanic floor is thrust back into the mantle and recycled.

Trench

Volcanic arc

Oceanic crust

Continental crust

Lithosphere →

← Lithosphere

Asthenosphere

◄ Percy Bridgman (1882–1961), pioneer of high-pressure research. The most abundant mineral in the Earth is named bridgmanite.

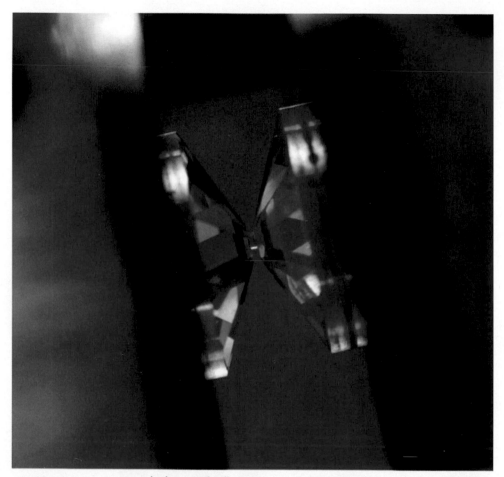

Inside this metal cell two opposing diamonds can create the pressures found deep within the Earth.

Map of the deep mantle. Once thought to be relatively featureless, modern science reveals it to be a very active region.

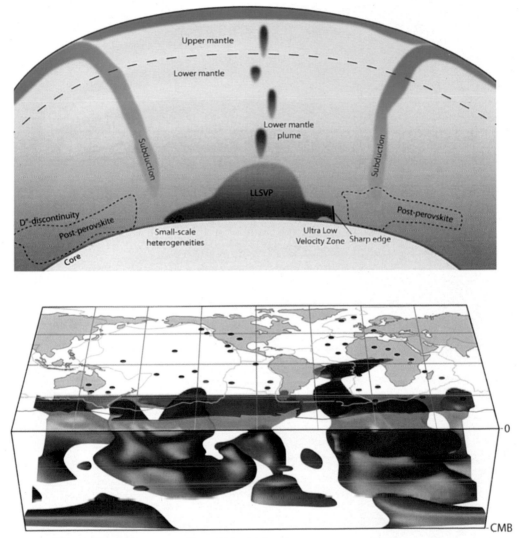

Vast structures, termed Large Low Sheer Velocity Provinces, lie atop the core–mantle boundary.

A rotating sphere of liquid sodium at the University of Maryland provides clues about the Earth's internal dynamo.

Ringwood and his team identified the significant minerals: majorite (a type of garnet found in the upper mantle of the earth), wadsleyite and ringwoodite (both high-pressure forms of olivine), all named for Australian researchers, including himself. Ringwoodite had been created in the lab and existed in the upper mantle but it had never been discovered on the surface of the Earth until a researcher looked at the composition of fragments of the 1879 Tenham meteorite and saw it contained ringwoodite, marking the first time that Ringwood's predicted mineral was found in nature. A few years ago a team from the University of Alberta isolated terrestrial ringwoodite in a brown diamond specimen formed in the mantle but later brought to the surface.

Using ever more intense pressures scientists homed in on an understanding. New atomic structures occur as a result of the rearrangement of atoms in olivine at a depth of 410 km, to form a closer crystal structure. The pressure at 660 km is about 24GPa, beyond anything available at the time. Below a depth of 660 km, evidence suggests that atoms rearrange yet again to form an even denser crystal structure, forming minerals called perovskite and magnesiowüstite. The 660-km seismic discontinuity also plays an important role in the dynamic state of the Earth's interior. It may be a barrier to whole mantle convection and the horizontal boundary between different convection cells in a layered convection model; there could be separate convection cells above and below it.

Mantles are found on other planets but none is like that in our planet. Among the solid planets, Mercury, Venus,

Earth and Mars, the Earth's mantle is the largest because we are the largest planet. We don't know a lot about Venus's mantle; a lot of what we know about its interior is based on analogy with the Earth, even though we know Venus behaves quite differently in many ways. When we look at its surface – using radar from orbiting probes, its thick cloud cover preventing any direct view – it appears to have a relatively young surface and there is no evidence of significant plate tectonics. It's unclear how its resurfacing works.

It's believed that the mantle of Mars will be very similar to the upper mantle of the Earth. Being a much smaller planet, its internal pressures do not reach the levels found in the Earth. The mantle on Mars should be moving very slowly, as the Earth's mantle does, but it seems to have a different style of convection. It has no tectonic plate activity. This is only found on Earth. Mars has produced some massive volcanoes, including the largest in the solar system – Olympus Mons. Some scientists believe the planet has what they call a stagnant lid, and that its mantle is an effective insulator keeping Mars's tiny iron core hot.

Returning to our own planet, we have seen that major insights were achieved using high-pressure apparatus combined with seismic data, but the pressures obtained were nowhere near strong enough to reach deep into the mantle. For that to happen a new approach was needed.

In his final years Alvin Van Valkenburg (1913–91) used to travel to science conferences and set up a device he had invented called a diamond anvil cell, and let anyone peer down his microscope into the high-pressure world. He would tell them with delight of his own first glimpse and how every day things changed completely in this new domain. Often younger scientists would thoughtlessly pass him by, thinking it was merely some form of amusement, unaware that some of the discoveries they were talking about had been made possible by the jovial man who let anyone look into his microscope. The great breakthrough was the diamond anvil cell, a device capable of incredibly high pressures rivalling those found at the very core of our planet. In a way the diamond anvil cell (DAC) and Alvin Van Valkenburg opened up more territory to explore than almost anyone else in the exploration of planet Earth.

He invented it while he was working at the US National Bureau of Standards. The DAC was based on a fact and a formula. Diamonds are the hardest, most rigid substance we can use. The diamond anvil cell consists of two gem-grade diamonds with their faces pressing against each other. A little metal ring gasket sits between them so as to create a tiny chamber that holds a sample measured in milligrams. It is an apparatus capable of lower-mantle pressures that is small enough to hold in your hand, because pressure is force divided by area, in this case a large force applied to a tiny area. This system has another great advantage: diamonds are transparent. We can observe samples directly, we can heat them up using lasers and we can aim x-rays through the diamonds to study the physical changes in the samples. The DAC is simple and quite remarkable. Valkenburg, along with Charles Weir and Ellis Lippincott, became the first human beings to watch matter transformed under a pressure of thousands of atmospheres. Initially, they got to 100,000 atmospheres, equivalent to 200 km in the Earth. 'We tried everything we could and broke diamonds left and right.' Then they reached 300,000 atmospheres.

Diamond cell anvils have become a very useful scientific tool enabling the structure of materials to be examined over a wide range of conditions. Atoms rearrange themselves under great pressure, revealing new materials. Under pressure the neat rows atoms like to form are disrupted. Often the new material created under pressure does not snap back to what it was. The structure is said to be metastable; diamonds are the best-known metastable material.

But other scientists showed little interest in Valkenburg's work. In the 1970s, he, along with Charles Weir and Ellis Lippincott, developed a new, more powerful diamond cell design. They all knew it was a good idea and that it would work. They filed a patent and set up a company, High Pressure Diamond Optics, and each put in a stake of $125. Sales were slow for the first few years and only Valkenburg stayed with it. Charles Weir retired and Ellis Lippincott died in 1974. But that year Lin-Gun Liu of the Australian National University used a diamond anvil cell to synthesise perovskite from a type of garnet and saw directly the mineral that, with its stability under pressure, comprises most of the mantle from the transition zone downwards. It inspired an enormous amount of research into the lower mantle. Liu found that silicate perovskite could form from all the major upper-mantle materials, olivine, pyroxine and garnet, and that was the cause of the 660 km discontinuity. The volume of the lower mantle exceeds 50 per cent of that of the Earth and silicate perovskite is the most abundant mineral in our planet, comprising 93 per cent of the mantle, yet we cannot find it at all at the surface of the Earth.

So there we have the composition of the mantle. Down to about 200 km it is composed of the minerals spinel and peridotite. Spinel consists of magnesium, aluminium and oxygen and peridotite is a coarse-grained rock consisting of the minerals olivine (magnesium or iron, silicon and oxygen) and pyroxene (calcium, sodium and iron). At a depth of 45 km and deeper these minerals change due to pressure when spinel reacts with pyroxine to create garnet peridotite (composed of olivine and pyroxine) that exists down to about 410 km. The mineral called wadsleyite – a high-pressure form of olivine – exists between 400 and 525 km depth, ringwoodite between 525 and 660 km and perovskite from 660 km down to almost the bottom of the mantle.

These three minerals ringwoodite, wadsleyite and perovskite determine much of the properties of the mantle. To honour Percy Bridgman's work in 2014, the Commission on New Minerals, Nomenclature and Classification of the International Mineralogical Association approved the name bridgmanite for silicate perovskite.

When I first encountered the device that can simulate the pressures deep within the Earth it was somewhat smaller than I had anticipated. When I thought of machines that deliver great pressure I thought of steam hammers bashing molten metal or the hydraulic rams used to crush cars. But the instrument in front of me was nothing like a muscular hydraulic press. It was so small I could place it on the palm of my hand. Yet in it resides a kind of magic. It transforms ordinary things. Oxygen, for instance, under tremendous pressure turns blue, then red and then becomes a shiny metal looking like steel. As we have seen, peanut butter turns to diamond. Then again, so does wood.

Using DACs, over the following years scientists increased the pressures they could obtain, being able to reproduce the conditions found at deeper and deeper regions of the Earth. In 1987 Elise Knittle and Raymond Jeanloz made high-pressure observations at 127GPa and concluded that silicate perovskite is stable throughout the lower mantle. In 1998 Sue Kesson and colleagues at the Australian National University, working at pressures of 135GPa, concluded, 'perovskite was found to be present and no additional phases were encountered'. Perovskite was regarded by most scientists as the base mineral of the mantle, the last transition of matter until we get to the Earth's outer core. How wrong they were!

13

Splinters of the Stars

We don't have to create the mantle in the laboratory to see it. Some things made in the mantle do find their way to the Earth's surface. Dan Frost of Bayreuth University told me that many volcanoes produce a basaltic lava full of what are called mantle xenoliths. Xenolith means foreign rock and they are bits of the mantle that have been ripped up and brought to the surface inside other rocks. If you open one of them you see they are comprised of olivine, which is a magnesium iron silicate mineral that has probably come from something of the order of 40 to 50 km down from a pressure of 1.5 GPa. He adds that we do have some xenoliths that come from deeper in the Earth.

In South Africa particularly you will find so-called Kimberlite eruptions. We don't know what these eruptions were like but we find that these volcanic pipes in the Earth that rapidly bring material from the mantle to the surface are full of xenoliths, and when we look at those rocks we find garnet, a red mineral with aluminium in it. They probably come from depths that go down to about 200 km. Also found in Kimberlite pipes are diamonds. The majority of diamonds, including all gemstones, are formed about 140 km down in the lithospheric mantle. Often these conditions are only met at the bases of old continents in their crystal basement rocks. It is thought that they start as carbon-rich fluids percolating through the rocks but at high pressure and relatively low temperatures, and in about a billion or two years they become diamonds.

Diamonds have been described as windows on the mantle and this is especially true for the so-called deep diamonds that are rarer than moon rocks. Fewer than a dozen ultra-deep diamonds have been found. In the past few years there was an important find of these elusive diamonds in Australia. White and just a few millimetres in size, they were discovered in the village of Eurelia in southern Australia. They formed much deeper in the mantle than normal diamonds, 670 km as opposed to 250 km within the mantle roots of ancient continental plates. Such diamonds are important because they are the only natural samples we can obtain of the lower mantle.

It is possible to analyse the carbon that the diamonds are made of and, remarkably, it has isotopic traces suggesting that the carbon the diamond is made of was once laid down as sedimentary deposits on the ocean floor and taken down into the mantle on a subducting slab! Other deep diamonds have been found in Brazil and Africa and researchers have noticed a pattern in that they are all found in areas that would once have been at the edge of the ancient supercontinent Gondwana. So the pattern emerges of subducting slabs descending beneath Gondwana carrying organic carbon which, over billions of years deep in the mantle, turns into diamonds. But the Eurelia diamonds seem to be much younger, perhaps 300 million years old, and so might influence where we look for diamonds in the future.

To describe a diamond as a cubic carbon mineral of simple composition with a remarkable range of properties is to miss the point completely. Like gold, diamonds are something more than material science, more than physics. They are the splinters of the stars, the tears of the gods. The word diamond comes from the Greek *adamas*, which means unconquerable. For centuries diamonds have been the most prized of gems and for the past forty years or so they have been studied to provide extraordinary information about the the their birthplace – the mantle.

14

D-Double-Prime

Perovskite has been well studied and few thought that under the pressure found at the base of the mantle it would change again. It seemed an ideal close-packed structure resistant to high pressures – the ultimate stable silicate – the major building block of planet Earth.

Using DACs today we can reach pressures almost of those at the Earth's centre (about 360GPa), but it is difficult. The diamonds deform at these pressures, along with their supporting metal parts. Because diamond is a superb conductor of heat, it's hard to keep a sample very hot. A 1984 lab accident involving a misdirected laser showed that we can actually melt diamonds. But that's undesirable, not just because it can wreck the equipment but also because diamonds still cost a lot of money. As it is, many anvils explode into dust, so experimenters have to leave the room during pressure trials. More research labs are using diamond cells, while solid-state physicists are putting frozen gases, water, industrial materials and exotic mixtures into the diamond cell to explore the basic properties of matter.

An amusing story is told about the scientists who use

the tiny samples held in DACs to determine large-scale properties of the Earth. In the early 1980s there was a talk by researchers who put iron metal in the diamond cell next to a few grains of silicate minerals. The news was that at deep-mantle conditions, the two substances reacted with each other. The speaker pointed to an enlargement on the screen and said, 'now, if we take the iron as a model of the core and the silicate as the mantle . . .' The roomful of scientists chuckled, because the whole sample was maybe a millimetre across!

Those who thought that perovskite was the final answer to lower-mantle composition were in for a shock. Around the turn of the century Kei Hirose of the Tokyo Institute of Technology and his colleagues started a systematic study of lower-mantle mineralogy using a laser-heated diamond cell. They compressed their sample and took it to SPring-8 – a giant synchrotron radiation laboratory in Japan. There an intense beam of x-rays was produced that was ideal for probing Hirose's squashed sample. The x-rays – light of short wavelength and high energy – travelled right into the heart of the crystal lattice structure of the sample, interacting with the rows of atoms in a way that depended upon how they are arranged. Reflected x-ray beams came off in all directions in what is called a diffraction pattern. By analysing this pattern scientists determined the way the atoms were arranged. SPring-8 is one of the best facilities in the world for this kind of work. Ten minutes' exposure to its x-ray beams can yield more data than 300 hours' exposure to some of the weaker x-ray sources used in the past. But when he analysed his sample, Hirose and his colleagues started to see things in the diffraction pattern they could not understand, so they decided to study simpler components separately. They saw that the diffraction pattern of perovskite changes completely between 110 and 120GPa.

The dramatic and to many totally unexpected announcement of the discovery of the perovskite to post-perovskite transition was made in April 2004 by Murakami, Hirose, Kawamura, Sata and Ohishi. Many scientists were astounded by the discovery and then wondered why it had taken so long to discover it. An interesting aspect of the discovery of post-perovskite is that a mineral phase with the same structure had been known for forty years but no one had thought it was a possible structure for the mantle. Post-perovskite – formed under the pressure found near the very base of the mantle – came along at just the right time to explain some very puzzling things that were seen down there.

The strange region at the base of the mantle was first identified by New Zealander Keith Edward Bullen (1901–76). He was by all accounts brilliant at mathematics as a child, often completing his homework in his head while riding his bike the five miles home. Then, as is often the case for earth scientists, an earthquake changed his life. In February 1931, when he was lecturer in mathematics at Auckland University College, the Hawke's Bay earthquake occurred which, with its death toll of 256, remains to this day New Zealand's deadliest natural disaster. It was caused by a slip in the subducting slab descending under New Zealand. 'Napier as a town has been wiped off the map.' Some coastal areas around Napier were lifted by 2 m. Forty square kilometres of sea floor became dry land. It all made quite an impression on the young Keith Bullen, stimulating a lifelong interest in seismology. Bullen spent 1931 through 1933 in England, at St John's College, Cambridge, where in his own words, 'I had the outstanding good fortune to be taken in hand by Sir Harold Jeffreys, who literally brought me down to Earth and rescued me from a purely mathematical fate.'

At that time the standard travel times used for the determination of the time of origin and of the location of earthquakes were those of Zoeppritz, as modified by H. H. Turner. It was known that there were errors in these tables of as much as twenty seconds. Calculating travel times is an iterative process, for the earthquake is located using one set of travel times, and the differences from that set of travel times are then used to determine a second set of travel times; the earthquake is then relocated using the second set of travel times, and a new location determined, and so on. Calculations of this kind were tedious and time-consuming, especially in the days of mechanical calculators. It was on tasks such as these that Bullen spent his years in Cambridge. Jeffreys said, 'Bullen's energy was phenomenal.'

As most people know, the Earth is not a sphere but an oblate spheroid, the polar and equatorial radii being 6,356 and 6,378 km respectively. In calculating the distance travelled by seismic waves from an earthquake this has to be taken into account. Travel-time tables as calculated for a sphere and those for an oblate spheroid are different, but not by that much, a few seconds. To advance, Bullen had to determine the variation of the Earth's density with radius. It is the discovery which established his reputation and became, as he said, 'a developing story which has fascinated the author over much of his working life'. Thus originated the Jeffreys–Bullen Seismological Tables in 1940, still used by the main centres of seismicity analysis. In the 1960s, however, some studies showed regional variations. At distances between 0 and 20 degrees from the epicentre of the earthquake the observed times differ from the J–B times by as much as six seconds.

Bullen returned to Auckland University College in 1934 and to Australia in 1940, where he was senior lecturer in mathematics at the University of Melbourne. It was there that he wrote his famous book, *An Introduction to the Theory of Seismology*. One of his great contributions was to build a mathematical model of the Earth's interior, something he started in the 1930s. All he knew at that time was that the Earth has a dense iron core surrounded by a thick rocky mantle. Without computers and using only the crude and cumbersome mechanical calculators of the time, he calculated the properties needed to explain the way the Earth bends seismic waves. He divided the planet into a series of shells labelled A to G, the crust being A and the inner core G. In his first model at the base of the lower mantle was the D layer. By 1950 he had decided that the data told him that the lower mantle was actually made of two very different layers, the lower one being a zone just a couple of hundred kilometres thick. So, in good mathematical fashion he divided D into D' and D"— D-prime and D-double-prime. The grand Earth divisions of the crust, upper mantle, transition zone, lower mantle and inner and outer core fit into his scheme. But that thin D' layer never got its own name, and in recent years it has gained quite a lot of attention.

The beginning of the testing of nuclear weapons meant that geophysicists did not always have to wait for earthquakes to be the creator of seismic waves. In 1954 there was a series of nuclear explosions at Bikini Atoll in the Marshall Islands. Father Burke-Gaffney of Riverview Observatory near Sydney saw the pulse-like arrivals of the seismic waves from the detonations. He and Bullen noted that the waves from the four blasts were separated by whole numbers of minutes and they said this was unlikely unless the shots had been fired at some well-defined time, for example, the beginning of the minute. They showed that the P travel times at Riverview and other stations were within a second or two of the J–B travel times. But, curiously, at three stations, Pretoria and Kimberley in South Africa and Tamanrasset in Algeria, the waves were all significantly early. The early precursors are now interpreted in terms of scattering at the core–mantle boundary (CMB).

The Bikini explosions took place while Bullen was president of the International Association of Seismology and Physics of the Earth's Interior and he was impressed by the potential of using atomic explosions for seismology. In 1955 he wrote to the president of the Royal Society of London, and to the Academies of Science in Washington and Moscow, proposing 'that for seismological and other experimental purposes one or more atom bombs be exploded during the International Geophysical Year'. He also used his presidential address to make a plea that information on explosions should be announced so that they could be used for scientific purposes. Before the meeting ended he received a telegram from the chairman of the US Atomic Energy Commission announcing a forthcoming detonation.

The mysterious seismic signature of the D" layer indicated to some that the deeper into the mantle you reached the more interesting it became. This region came into focus gradually as information from seismography combined with new wave-analysis techniques, geodynamics and geochemistry was gathered to form a new view. The first hints came from seismology when it was suggested that the lower few hundred kilometres of the mantle had a reduced seismic velocity gradient. By the early 1980s modifications to this view were made. There was definitely structure down there. An increase in velocity was seen between 250 and 150 km above the CMB. D-double-prime was slowing emerging from the depths.

The rocks at the base of the mantle are squeezed at 135 GPa and are white-hot. The scientific picture of D" is ambiguous: different people see it different ways but almost all recognise its significance. One popular idea is that it is the place where slabs from the lithosphere come to rest having made the journey all the way from the surface and where iron–silicate slag accumulates at the edge of the outer core. Others conjecture that D" is involved in the release of energy that sends material back up to the surface, like a lava-lamp the size of a planet!

15

Dark Matter

When was the last time you admired a jade pendant or necklace? Do you know the remarkable story of where these deep-green semi-precious stones come from? As the slab descends the rocks from the ocean floor begin to change under the increased heat and pressure and some of them, those rich in sodium and aluminium and poor in iron, are metamorphosed in a particular way. The pressure is high but the slab remains cooler than the surrounding mantle and that enables the formation of jadeite. Sometimes fragments the size of cities break away from the slab. The newly formed jadeite in them survives and is entrained upwards as the mantle rock melts and rises. Tens of millions of years later, the semi-precious stone is found in molten rocks on the flanks of volcanoes. Today jadeite is found in only twelve places on Earth. The ancient Chinese called jadeite 'the stone of heaven'. How wrong they were.

Returning to our ride along with the oceanic slabs we see they have now reached the transition zone. In a review of images of subducted slabs around the Pacific one group of scientists noticed that some slabs tend to be trapped at a depth of 660 km. Ed Garnero: 'It hits the mantle transition zone where the atoms reorganise to a closer-packed configuration. It's denser and it's more viscous as well. So the slabs will often broaden a little bit. Sometimes it seems as if they lay down on that boundary. Some keep falling.'

Some believe that the slabs that are descending the most steeply penetrate the 660-km barrier and that those encountering it at a shallow angle often stagnate there and move horizontally. Some of them will resume their downward plunge after a period of horizontal motion along the 660-km-depth barrier. Some research suggests that the younger slabs can penetrate the barrier while older ones can stall. Certainly if you look at seismic images of subduction zones you can see the disturbance in their downward trajectory caused by the 660 and sometimes by the 440. It's a behaviour that may be linked to changes in the rocks in the slab. As we have seen, minerals change to a more compressed form due to the pressure and this increases the density of the slab, sometimes by as much as 16 per cent, which can then resume its downward journey.

Beneath 660 km, there's not much in the seismic picture until the base of the mantle. Following the slab down during its slow but inevitable journey deeper into the heart of our planet takes us through the middle of the mantle, which is a region where the changes are slower than those we have seen above it. But that quiescence is temporary, for we are soon to approach one of the most extraordinary, significant and little understood regions of our planet. Ed Garnero: 'Everything underneath the slab is being pushed aside. There is a pressure in the mantle caused by the descending slab so stuff is getting out of the way or stuff is moving with.' The slab is heading for D" and the core–mantle boundary and it's going to find something remarkable.

'The core–mantle boundary', he adds, 'is one of the strangest and most significant places in Earth. It contains what could be termed the Earth's Dark Matter.'

16

Borderlands

Adam Dziewonski (b. 1936) was born in Lwów, then in Poland but now a part of Ukraine, and is a Polish-American geophysicist who has made seminal contributions to the determination of the large-scale structure of the Earth's interior and the nature of earthquakes using seismological methods. Along with Don Anderson of Caltech he pioneered a new technique for looking into the depths of the Earth. 'I am a geophysicist who has spent most of his professional career mapping the deep interior of the Earth. Nearly twenty years ago I developed a method which would allow us to see in three dimensions features that are as deep as 1,600 km beneath the Earth's surface. This approach is now called "seismic tomography", because of the analogy with medical tomography. Mapping of regions that are anomalously hot or cold is likely to tell us the origin of the driving forces of plate tectonics.'

At the start of my scientific career I was a radio astronomer at Jodrell Bank near Manchester, England, using the giant telescopes there. One way to use them is to connect them together and take advantage of all of them looking at the same cosmic source at the same time. The signals received by each could be combined to synthesise the view that would be seen by a telescope much larger than any individual scope – a telescope, in fact, the size of the distance between them – and they could be at either ends of the earth. It was a supreme example of signals processing. Seismic tomography is essentially the same technique. Look at all the data from the seismic waves generated by earthquakes and try to develop a three-dimensional image of the Earth that satisfies all the data, all the fast- and slow-velocity regions. What has been found is a new view of the inside of our planet, as profound as it is sensational.

Adam Dziewonski remembers when he got the idea of seismic tomography: 'My interest in global seismology started sometime in 1967 when I heard Frank Press talk about using methods to select from some three million randomly generated models of the Earth only five that satisfied the observations. Even those five models were quite different and it seemed to me that we ought to do better.' He knew the key was better data. The Alaskan earthquake of 1964 was recorded by the new global network of seismometers but it was recorded on photographic paper, so it was impossible to put them into the computers of the time. This led to a two-year programme to convert these records into strings of numbers that could be analysed in a computer. It was a gamble that paid off.

The information was in the seismic data and the world's greatest collection of such data is held in an industrial block in Kennet, a suburb of Newbury in southern England. Wayne Richardson, a New Zealander who is as expert as anyone in reading the secrets of the P and S waves, showed me around the International Seismological Centre which monitors and collates every bit of data available about seismic waves travelling through the Earth. Every day new data come in by email to be registered and evaluated, producing a definitive record of the waves that travel through the Earth and their origins and arrival times at almost every station on our planet. In the early days it contained reports of arrival times from some 1,000 globally distributed stations for, roughly, 10,000 earthquakes a year, but it has since grown. He showed me their storeroom, a warehouse of bulletins, books and periodicals all containing earthquake data. Old records, paper notes and correspondence awaiting transfer into digital format. There were boxes everywhere containing records from almost every country on earth going back decades: Russia, India, Argentina, Iran, Botswana to name but a few. There is no room anywhere on Earth like that at the Seismological Centre at Kennet and its repository of the shaking Earth.

Looking at these records something initially puzzling is apparent. After the giant Alaskan earthquake of 1964 there was a forty-year lull in giant quakes followed by a resurgence. Since 2004, devastating quakes have rocked Sumatra, Chile, Haiti and Japan, leading to speculation that we might be living in an age of great earthquakes, similar to a global cluster seen in the early 1960s. Some researchers have even suggested that large quakes possibly trigger each other. To see if the clusters of the 1960s and 2000s were significant, researchers examined the timing of the earthquakes with magnitude 8.3 and above during the past 100 years.

The timing seen between real-life large quakes was what might be expected from randomness. Curiously, some research does appear to show that smaller earthquakes seem to communicate over global distances. After big quakes, there are lots of micro-tremors all over the planet, but for some reason they don't seem to grow into big earthquakes. We clearly need to learn more about why some earthquakes grow big and why some don't.

There is evidence that big earthquakes might actually calm the Earth. The 11 April 2012 Indian Ocean earthquake was the largest strike-slip – horizontal moving fault – ever recorded. The magnitude 8.6 quake triggered earthquakes worldwide for up to six days, according to the US Geological Survey. But once the triggered quakes stopped, scientists were surprised to find a sharp drop in moderate earthquakes for more than three months. There were no earthquakes bigger than magnitude 6.5 for ninety-five days. Normally, such quakes occur every ten days. The chance of going ninety-five days without one is less than about 1 in 10,000. It's thought that the Indian Ocean earthquake's unusually energetic seismic waves could have reduced stresses on faraway faults, delaying earthquakes.

Dziewonski's idea was that the travel-time data for many ray paths, criss-crossing the Earth between various points near the Earth's surface and reaching different depths in its interior, could be resolved into a three-dimensional model.

This is 'seismic tomography', and it conceptually resembles the medical CAT scan. The early results were reported at conferences in 1974 and 1975 and scientists eagerly awaited formal publication in a scientific journal. Then a single issue of the *Journal of Geophysical Research* changed the way we look inside the Earth.

Dziewonski and Anderson found four 'grand' structures down there. They include two regions of higher than average wave speed, inferred to be cold and sinking mantle, one under the western edge of the Americas and the other under southern Eurasia. Looking at speeds and density there were two large regions with low sheer wave speeds and higher than average density lying beneath Africa and the Pacific. The African region towers above the core–mantle boundary (CMB) and the Pacific region is somewhat lower. Together they cover half of the CMB. Technically they are known as Large Low Shear Velocity Provinces (LLSVPs) but they have been given names, Tuzo and Jason, after pioneering earth scientists W. Jason Morgan and Tuzo Wilson.

Tuzo and Jason are truly giant structures each 15,000 km across and rising 500–1,000 km above the CMB – our planet's underground continents. Recent research suggests they are ancient and are likely to have formed 4.4 billion years ago when the Earth was young. They alter the passage of seismic waves because they have a different composition and temperature than other material at the base of the mantle. Alongside and around them is the D" discontinuity. Some think of it as a landscape where the mountainous Tuzo and Jason are surrounded at their flanks by zones of post-perovskite. Around the edges of the LLSVPs are intermittent layers of 5–40 km of mantle rocks. These are called Ultra Low Velocity Zones (ULVZs) because of their effect on seismic waves. In some seismic reflections from the region there are hints of small structures perhaps only 10 km or so across; some speculate they could be zones of partial melting or perhaps even the remains of subducted slabs. Recent data obtained by the Earthscope USArray – a grid of more than 400 seismometers placed across the United States – on the CMB has seen ridges of mantle rocks reaching up to 100 km above the CMB. The edges of Tuzo and Jason appear sharp and scientists speculate that they deflect material upwards, for what goes down to the CMB can come back up again.

17

Plumes

That there is some recycling of material within the Earth is undisputed, but there is much debate about its details and how long it takes. If slabs are subducted into the Earth then something must come back up again, but from where? Is it just from the surface at mid-oceanic ridges, just below the surface, as in volcanoes, or does material come from somewhere deeper? We know what goes down but is anything coming back up?

Looking at rocks that erupted twenty million years ago on an island in the South Pacific called Mangaia, the southernmost of Polynesia's Cook Islands, shows that the molten rock just beneath the crust varies in composition from place to place and one interpretation is that crust that once resided on the surface has altered the composition of the mantle in an uneven way. Closer inspection of the mantle in this region reveals something even more fascinating. It has traces of material that was on the surface of the Earth some two and a half billion years ago, before photosynthetic organisms filled the atmosphere with oxygen. Analysis of the Hawaiian volcano Mauna Loa shows that its lava contains material that once comprised sediments dragged into the Earth on an ancient subducting slab. That is the case in many regions. Rocks are recycled, but how? Slabs may reach their graveyard at the base of the mantle but then their material is sent back up again, and some believe this is done by so-called superplumes.

Iceland is perhaps the most volcanic country on Earth as it straddles two tectonic plates separated by a spreading centre. But its intense volcanism is greater than would be expected just from being over a spreading zone, leading to the suggestion that hot rock is reaching it from greater depths. Some have suggested it has a rising superplume beneath it.

The thought is that something happens at the sharp edges of Tuzo and Wilson, something that could force rock to rise through the lower mantle and perhaps continue to the surface. One of the clues that this could be happening comes from volcanoes. As we have seen, about 95 per cent of the world's volcanoes are located near the boundaries of tectonic plates, and they are there because the descending slab oozes water that reduces the rock's melting point, causing it to rise to the surface and form a volcano, but there are other types of volcanoes. The other 5 per cent are thought to be associated with so-called mantle plumes and hot spots. Mantle plumes are areas where heat and/or rocks in the mantle are rising towards the surface. A hot spot is where it reaches the surface. There is strong debate in the geophysics community about the reality of mantle plumes; like so much in that fascinating science, things are not settled.

In the 1960s, Tuzo Wilson (1908–93) noticed something

remarkable about ocean islands. On a map of the Pacific basin, he found three linear chains of volcanoes and submarine volcanoes (seamounts). They were dispersed across thousands of miles but what caught his eye was that they were parallel to each other. He collated all the available data on them and an interesting pattern emerged. For each chain, the islands become progressively younger to the south-east and the extreme south-east end of each chain was marked by active volcanoes! At first sight the Hawaii–Emperor volcanic chain might remind you of the volcanoes we have encountered near a subduction zone. But the volcanoes of the Hawaii-Emperor volcanic chain get progressively younger towards Hawaii. Perhaps there was a hot spot in the Earth's mantle that Hawaii, on its moving tectonic plate, was passing over, causing volcanoes to form above it and fade away when Hawaii had moved on.

By the early 1970s the presence of hot spots was accepted. In 1971, W. Jason Morgan (b. 1935) suggested a more important role for hot spots. He speculated that they result from hot, narrow plumes of material that rise from deep within the mantle. As it reaches the base of the lithosphere, it spreads out, a little like a blowlamp. Morgan proposed twenty different hot spots, some located along mid-ocean ridges and others, like Hawaii and Yellowstone, located within tectonic plates. Researchers think these mantle plumes are shaped like mushrooms: narrow streams of molten rock capped by bulbous heads.

Not everyone believes the mantle plumes hypothesis. Massachusetts Institute of Technology seismologist Qin

Cao and her colleagues have used seismic waves to peer at what's happening directly below Hawaii and they maintain they cannot detect the head of a mantle plume; rather, they see what they call a thermal anomaly about 800–2,000 km wide that powers its volcanoes. They suggest they are there because of a vast pool of hot matter on top of the lower mantle and not from its bottom.

Hawaii sits on the world's most vigorous hot spot, but if it is a plume it is a complicated beast. Looking at it as closely as possible with the seismic data shows an unexpected bulge in what might be a plume some way below the bottom of the lithosphere and much deeper than expected. By way of an explanation it has been suggested that if the plume is rich in a mineral called eclogite, which is denser than typical mantle materials, that means it will stall at a depth of about 400 km, where it should spread out horizontally. Eventually the eclogite-rich rock would become buoyant enough to rise as a thin plume. It's a possibility, but in the plume business no one is exactly sure. Beneath Hawaii some scientists have detected what they think is a warm pool about 110–155 km below the surface, but it seems to be centred 100 km west of the main island instead of directly beneath it. Perhaps the Hawaiian plume bends as it approaches the surface.

A Franco-German experiment known as RHUM-RUM (Réunion Hotspot and Upper Mantle–Réunions Unterer Mantel) is also looking at the place where a plume has been suggested, this time the area underneath the island of La Réunion in the Indian Ocean. La Réunion has the advantage of being close to Madagascar and relatively near to southern Africa, where researchers can place seismic instruments more easily. They deployed nearly sixty seismometers on the ocean floor over 4 million square km as well as another thirty instruments on land, making this project the largest yet in the search for a mantle plume. They aim to image a whole section of the mantle from the crust to the core. Years of data have been obtained but for now the analysis continues.

The sizes and occurrence of mushroom mantle plumes can be predicted using theories of fluid motion, with plumes of about 2,000 km in diameter taking about 900 million years to rise from the CMB to just below the surface. The number of mantle plumes is predicted to be about seventeen. The plume hypothesis has also been studied since it was first proposed using laboratory experiments conducted in small fluid-filled tanks, and plumes produced in that way were used as models for the much larger postulated mantle plumes. On the basis of these experiments, they are expected to be made of two parts: a long thin conduit connecting the top of the plume to its base, and a bulbous head that expands in size as the plume rises. The bulbous head forms because hot material moves upwards faster than the plume rises. In the late 1980s and early 1990s, experiments with thermal models showed that, as the bulbous head expands, it might carry with it some of the adjacent mantle.

When a plume reaches the base of the lithosphere, it is expected to flatten against this barrier and to undergo decompression and melting. Large volumes of basalt magma are formed which may then erupt on to the surface, and there are indications that when this happens the effect can be devastating.

The Deccan Traps on the Deccan Plateau of India are one of the largest volcanic features on Earth. The word trap comes from the Swedish *trapp* or *trappa*, meaning stairs, and indicates the step-like hills of the region. The traps cover half a million square kilometres, are more than 2 km thick and have a volume of half a million cubic kilometres. Estimates suggest that when it erupted it was several times that – spewing on to the surface of the Earth. Imagine it: half the size of India covered in magma, a 30,000-year devastation that took place between sixty and sixty-eight million years ago. The release of vast amounts of volcanic gases, particularly sulphur dioxide, would have caused substantial climate change. Some models predict a global temperature drop of 2 degrees C. That's what some think can happen sometimes when a mantle plume erupts. The plume that caused such devastation is thought to come from the Réunion hot spot, the region currently under such intensive investigation. The mantle plume that may lie under La Réunion in the Indian Ocean has apparently burned a track of volcanic activity that reaches about 5,500 km northward to the Deccan.

It's not just the Deccan Traps. There are similar regions in Siberia, the Karoo-Ferrar basalts in South Africa and Antarctica, the Paraná and Etendeka traps in South America and South-West Africa respectively (before they separated due to the opening of the South Atlantic Ocean), and the Columbia River basalts of North America. Flood basalts in the oceans are known as oceanic plateaux, and include the Ontong Java Plateau of the western Pacific Ocean and the Kerguelen Plateau of the Indian Ocean. Such disasters are widespread and on geological timescales common. Eleven of these eruptions may have occurred in the past 250 million years, many coinciding with mass extinctions.

The key to solving the plume problem is, as usual, better data and that means placing more instruments on the surface. It is an approach that has worked in places such as Yellowstone National Park in Wyoming, thought to lie atop a mantle plume responsible for some of the biggest volcanic events in the past few million years. Those data were gleaned from a project called EarthScope that has been moving across the United States for the past decade deploying temporary seismometers in a 70-km-square grid. It clearly saw a structure called Yellowstone plume 3, showing it to be a hot, narrow upwelling that comes from at least 900 km deep in the mantle.

The big question is: can we see mass extinction events on the way up? Some scientists believe we can by looking for the plumes. Such a thing is seen in the south-west Pacific near the Fiji Tonga subduction zone. It's 700 km deep, has a structure consistent with a massive temperature anomaly and may be rising. It could render the Earth uninhabitable for humans and it will reach the surface in an estimated 200 million years.

We can see this movement and feel it using sensitive equipment. Sail towards the centre of the Indian Ocean and you will lose weight, only very, very slightly because the Earth's pull of gravity is slightly less there than at other places on its surface. Similar reductions in the

Earth's gravity can be detected with sensitive apparatus in the north-east Pacific and the Ross Sea. The reason for the slight decrease may be due to descending oceanic slabs, dense structures moving downwards and further away from the surface. But there appear to be even more dents in the Earth's gravitational field due to plumes of less dense material rising upwards.

18

'The old boy beamed upon me'

In 1862 the distinguished physicist William Thomson calculated the age of the Earth. He believed that it had started out very hot and had cooled since its formation: 'This earth, certainly a moderate number of millions of years ago, was a red-hot globe.' His 1864 estimate, based on how long a solid takes to cool, was from 20 to 400 million years old. In 1897 Thomson, now Lord Kelvin, settled on twenty to forty million years old. It was too short a time for evolution and also for the geologists. His former assistant John Perry published a paper in 1895 based on an Earth that wasn't a solid but transferred heat by convection and obtained an age of a billion years, much better for the biologists and the geologists, but his work had little impact.

The discovery in 1903 of radioactivity led to Ernest Rutherford famously making the case for an older Earth at a lecture attended by Kelvin himself. Later Rutherford wrote: 'Behold! the old boy beamed upon me.'

Radioactivity is an essential part of the Earth that provides about half the heat our planet gives off, and one by-product of the process are neutrinos. These are remarkable particles that hardly interact with matter at all. They were first postulated to tidy up missing energy in radioactivity, and were discovered in 1956. They are produced in certain types of radioactive decay and in some nuclear reactions, especially at the centre of the Sun. The heat energy produced by the nuclear furnaces at the centre of the Sun takes half a million years to get out, but the neutrinos produced in the nuclear reactions get out immediately, flashing away and not even noticing that the Sun is there. We have neutrino maps of them emerging, proving to us that the Sun is, right now, producing energy. Neutrinos are everywhere: sixty-five billion of them from the Sun pass through every square centimetre of the side of you facing the Sun, and they do you no harm. We have detected them from the Sun, from an exploding star in an outrigger galaxy to our Milky Way and from the depths of the Earth.

Neutrinos in the Earth are produced by the radioactive decay of potassium, thorium and uranium and they account for 99 per cent of all heat generated by radioactivity from inside the Earth. But could they ever be used to probe the Earth? In 1984 Lawrence Krauss, Sheldon Glashow and David Schramm published a paper presenting estimates of the geoneutrino flux as well as comments on their detection. In 2005 they were actually detected using the Japanese Kamioka Liquid Scintillator Antineutrino Detector (KamLAND) that picks up neutrinos passing through a tank of 1,000 tonnes of mineral oil! The whole device is in a mineshaft. As a certain kind of neutrino travels through the mineral oil it causes a flash of light which is picked up by banks of light sensors placed around the tank. The tank is 13 m. in size and, as we have seen, sixty-five billion neutrinos pass through every square centimetre. In one experiment running for 145 days it detected just fifty-four of them but it was enough to draw meaningful conclusions. In 2005 an observing run of 749 days picked up 152 neutrinos, of which twenty-eight were from within the Earth. In 2011 the data was updated and in 2,135 days 841 neutrinos were picked up, of which 106 were terrestrial. In 2010 another neutrino detector in Italy detected neutrinos from the Earth.

Some scientists believe that neutrinos could be a new tool for probing the Earth, and certainly such data would be useful in answering many interesting questions about our planet's origin, composition and heat budget. A neutrino observatory designed to study the Earth would first have to get away from the continental crust because, although the crust comprises only 0.34 per cent of the mass of the Earth, it harbours 40 per cent of our planet's radioactivity. This means the neutrino detector would have to be on the sea floor, probably near Hawaii, as far away from continental crust as it is possible to get. It's estimated that the flux of neutrinos coming out of the Earth is something like a million per square centimetre – far, far less than the number coming from the sun. This means that in the future a neutrino detector positioned on the sea floor would not only have to be big in order to capture even a minuscule fraction of the neutrinos passing through it, but also be able to tell from which direction they are coming. It's a fascinating idea, but I don't think it will happen very soon.

We have seen that the mantle is no bland space separating the active crust from the extraordinary core. It has a vital function and rhythm of its own that regulates long-term activity on the Earth's surface, providing the driving force behind the movement of the continents in their half-a-billion-year cycle of supercontinent formation and dissolution. Material is cycled from the surface down to the base of the mantle and back up again. It has mysterious continent-sized structures within it and it plays a role in establishing the conditions needed for life to develop and survive.

On our imaginary journey we have reached the base of the mantle. We have travelled about 2,950 km through the crust and the mantle but haven't yet reached the halfway point from the crust to the core. More than 3,300 km remain to be traversed and we are about to leave our planet of rock behind and experience the most dramatic boundary and change of scenery that our planet has to offer. An extraordinary inner world the size of the planet Mars awaits us.

19

The Protector

'And looking around him he saw a giant ocean as far as the eye could see . . . he soon realised they had reached a world within a world'

Jules Verne

The next region we must traverse on our journey is the outer core, a place like nowhere else we have encountered on our descent through rock. It's made of liquid metal moving in currents and swirls confined between the inner core and the mantle, but its influence extends further than the Earth itself.

A million and a half kilometres closer to the Sun from the Earth are three active satellites stationed in a stable orbit that enables them to monitor the Sun. It's called a Lagrange point and it's where gravitational forces balance, allowing a satellite to remain in a relatively fixed position without too much expenditure of fuel. One of them, called SOHO, or the Solar and Heliospheric Observatory, has been there since 1996 keeping an eye on the surface of the Sun and what it might throw at us. Along with WIND, a satellite that arrived at its position in 2004, and the Advanced Composition Explorer, designed to study the particle wind from the Sun, they keep watch on the sometimes angry Sun, sending warnings when clouds of electrically charged particles are coming our way so that we can prepare for them. Such outbursts can cause problems on Earth, as was the case in 1859 and 1989.

The solar wind and occasional larger disturbances race past the three sentinels towards the Earth's defences, for surrounding our planet at great distances is a magnetic shield that protects us against most of what comes from the Sun, and from elsewhere in space. If this shield were not there life on Earth would not be possible. Its surface would be subject to intense bursts of radiation from solar flares. A magnetic sheath does not protect the Moon and any astronaut out in the open on a particular day in 2005 would have received a lethal dose of radiation. In addition, the continuous stream of particles from the Sun – the solar wind – would over time have stripped the Earth of its atmosphere were it not for the magnetic protection. Mars lost its magnetic shield early on in its lifetime and has had its atmosphere almost completely stripped away by the solar wind, probably rendering it lifeless. Were it not for the Earth's protection – the magnetosphere – it would have suffered the same fate.

The magnetosphere stretches 64,000 km into space towards the Sun to the so-called bow shock – where the onslaught from the Sun is partly held back – to over six million kilometres in the opposite direction down the extended magnetotail. It seems to have been in place for at least 3.4 billion years. But perhaps the most remarkable fact of this life-preserving shell is that it is made inside the Earth and reaches out beyond its rocks and atmosphere from the metallic currents in the Earth's outer core. We are creatures of the surface, of light, of carbon and energy, and we are also children of the core.

As we have said, seismology is by far the main way we have found out about the structure of the Earth beneath us, but there are other ways. One is to study the influence that the Sun and Moon have on the Earth in the form of tides. Tides are not restricted to water; as the Moon and Sun seem to move in the sky their gravity pulls the Earth towards them, making it become egg shaped. The Earth, being rigid, will resist and studying its deformation can provide important information about its strength, especially the nature of the upper mantle. Beno Gutenberg worked out six methods for determining the Earth's strength and concluded that it was highly rigid. Little-known Russian scientist Leonid Leybenzon (1879–1951) was the first to make the suggestion for a non-rigid or liquid core. He calculated that the thickness of the mantle was between 30 and 50 per cent of the Earth's radius. But his 1911 paper attracted little attention. We have seen that it was Richard Dixon Oldham who first put together the seismic data to determine the existence of the core of the Earth when he noticed that there was a shadow in the distribution of S waves beyond a radial distance of 120 degrees from an earthquake, although he was not absolutely certain, as there were some seismic waves in the shadow zone that he couldn't easily explain by his core hypothesis. Then Beno Gutenberg, who took a long time to come round to the idea that the Earth's core could be liquid, calculated that there was a very sharp change in seismic velocity 2,900 km below the surface – about halfway to the centre. In 1914 Oldham revived the possibility that the waves seen in his shadow region were not S waves but reflections from the outer core's surface. Perhaps it was liquid. No one was convinced.

It was not the first time there had been a debate about whether the central regions of the Earth were solid or liquid, although it was the first time there had been good data. In fact there was a considerable scientific battle in the eighteenth century over whether the Earth was entirely solid or had a liquid core, and the solidists won. In 1909 the president of the Geological Society of London declared: 'Thus, through the shifting sands of an ancient and prolonged controversy, terra firma, indeed terra firmissima, has at length been reached.' Lord Kelvin's view ruled when he proclaimed that 'the Earth is as rigid as steel'. At the end of the nineteenth century there was something of a stalemate: British and American scientists regarded the Earth as solid, German scientists were not so sure. Wiechert, for example, did not believe that compression by the weight of overlying rocks would account for the density of the inside of the Earth; it had to be made of a denser substance, but what that was, and what form it took, he was not sure.

Despite Wiechert's views most seismologists thought that this change in velocity 2,900 km below the Earth's surface did not relate to a change to liquid. For them the entire Earth was solid and the change was due to the increasing presence of iron. There were several lines of evidence that suggested that the Earth was entirely solid. It was well known to physicists at the time that increasing pressure increased the melting point of many substances, so the high pressure inside the Earth could keep whatever it was made of solid despite its high temperature. The discovery of radioactivity also suggested to some that the inside of the Earth need not get hotter the deeper you went. The heat just below the surface of the crust, they argued, might be due to radioactivity and not a general trend of 'the deeper the hotter'. But while all of these arguments were being exchanged the secret of the inside of the Earth was in the seismic data, although interpreting it was difficult given preconceptions. Oldham's interpretation that S waves did not travel through the core was not definitive and Gutenberg was certain that the core was solid. By about 1925 it was clear that S waves did not travel through the core but most seismologists still believed the core was solid. A large part of the solution to the stalemate came from Harold Jeffreys in 1926 in his paper 'The Rigidity of the Earth's Central Core'. His argument was not based upon whether or not S waves travelled through the core. The mantle, he said, was more rigid than the average rigidity of the Earth as a whole as determined by the tides raised in the solid Earth by the Sun and the Moon. So what is beneath the mantle must be of lower rigidity and that, coupled with the S-wave data, provided a strong argument for it being liquid But is it really iron? You could almost hear the seismologists' sighs of relief.

Once again the evidence was there but it was mixed with diversions. In the late eighteenth century it became possible to calculate the total mass of the Earth and it came out to an average density about 5.5 that of water, that is, twice the density of surface rocks. Iron is the most abundant element known to have a high enough density and also many meteorites were iron, so it was reasonable to think that there could be a lot of iron down there. Before the first principles of the structure of the atom were determined just over a hundred years ago no one really knew how much matter could be squeezed. But after Ernest Rutherford's model of the atom in 1911 some believed that matter could be squeezed more than had been thought. Astronomers quickly suggested that this could happen in stars and then they showed that the sun is made mostly of hydrogen and that iron is only present in small amounts. So if the Earth is formed from the same material as the Sun's atmosphere, perhaps there is not that much iron at the Earth's core. Indeed, in 1941 Werner Kuhn (1899–1963) and Alfred Rittmann (1893–1980), both Swiss, suggested that the Earth contains a lot of hydrogen. In 1948 William Ramsey suggested the Earth formed from material similar to the crust and there would be a dense liquid inside, but not necessarily much iron. But the theory of an iron heart to the Earth would not go away as every other idea faded. Iron seemed the only explanation, and when it was accepted further calculations showed that iron was actually too dense and had to be combined with a lighter element, such as nickel, to make the numbers work. It was not until 1957 – three decades after Jeffreys's work – that Gutenberg could bring himself to state publicly that the core was liquid.

The greatest boundary we will cross on our journey from the crust of the Earth to its centre is not that between the atmosphere and surface rocks, however much of a jump that is. It is the transition from the lower mantle to the outer core, from solid to liquid. Suppose in our imagined capsule we traverse this boundary; would we find it abrupt? How sharp the transition is a matter for debate. The liquid iron of the inner core is highly reactive with the mantle's silicate rocks and it is expected that some liquid iron will percolate into the mantle. It might react with silicon and oxygen as some high-pressure experiments indicate. If it did this material would be heavier than the surrounding mantle material and will therefore stay at its base. Some would be incorporated back into the mantle but some would stay and perhaps be swept up to the Large Low Shear Velocity Provinces, forming the Ultra Low Velocity Zones we have already encountered.

All this means that the transition between the mantle and the core could be a little mushy, although we don't know how deep it would be. In our capsule we could station-keep just above the boundary in the white-hot rock of the lower mantle, and then move perhaps a few tens of metres and be in the liquid metal of the outer core. Imagine you could see through this molten iron as you move a little further into the upper core. Looking back the way you came you would see you are inside a giant sphere of rock. Close up there is no indication that it is the inside of a giant sphere of radius 3,481 km; it would appear as a great wall. The core of the Earth is larger than the planet Mars and far more alien. It has one-sixth of the volume of the Earth yet one-third of its mass, and it is liquid, dense yet not thick. If you donned super-protective gloves you could run your hands through it like water.

It is this liquid – molten iron, nickel and a few other elements – that profoundly affects the nature of our planet and protects us from the harshness of the cosmos. In the mantle we suspect that there may be aspects of the sub-duction cycle that are important for life on the surface. But in the liquid core we have no such doubts. We are certain life on our planet could not have survived without it, for out of its liquid motions emerges our great protector – the Earth's magnetic field.

20

Magnetic Dreams

When Edmond Halley is mentioned most people think of the famous comet named after him. But the comet whose return he predicted is far from his most important scientific achievement. He carried out the first truly scientific expedition sent out by Britain into the far world and he was sailing under royal warrant

> Whereas his Majesty has been pleased to lend his Pink [sailing ship] the Paramour for Your proceeding with her on an Expedition, to improve the knowledge of the Longitude and variations of the Compasse, which Shipp is now compleatly Man'd, Stored and Victualled at his Mats. Charge for the said Expedition; You are therefore hereby required and directed, forthwith to proceed with her according to the following Instructions.
>
> You are to make the best of Your way to the Southward of the Equator, and there to observe on the East Coast of South America, and the West Coast of Affrica, the variations of the Compasse, with all the accuracy You can, as also the true Scituation both in Longitude and Latitude of the Ports where You arrive.

His first expedition started in November 1688 but trouble with an insubordinate lieutenant who did not recognise Halley's authority as captain caused him to return from the West Indies the following summer, without having crossed the equator. He set off again in September 1699.

Whereas his Majesty has been pleased to lend his Pink ye Paramour, for your proceeding a second time with her on an Expedition to Improve ye knowledge of the Longitude & variation of ye Compass, which ship is now Compleatly mann'd, stored and victualled at his Majesty's Charge for ye said Expedition, you are therefore hereby required and directed forthwith to proceed with her according to ye following Instructions.

You are without loss of time to Sett Saile with her, and proceed to make a Discovery of ye unknowne South lands between ye Magellan Streights & ye Cape of good hope, between ye Latt:d of 50 & 55 South, if you meete not with ye land sooner observing ye variation of ye Compass with all ye accuracy you can, as also ye True Scituation both in Longitude & Latt:d of ye ports where you arrive.

Mapping the 'variations of the Compasse' was his primary goal, for all sailors who used these compasses knew they often behaved strangely at sea.

Magnetism: even today it evokes a feeling of strangeness, as if it is a force that seems to defy common sense, as one attempts to place like poles together and feels the resistance. The first puzzle is that magnets always occur in the form of dipoles; there is always a North pole and a

South pole. It is possible to have electric dipoles, but they consist of a positive charge and a negative charge. It's possible to separate them and study each one separately. The same is not true for magnetic dipoles; trying to separate the two poles of a magnet results in two dipole magnets, each with its own North and South pole. No one has been able to isolate a North magnetic pole or a South magnetic pole. Magnetic monopoles, if they exist, have not yet been detected.

For millennia mankind has been both fascinated and puzzled by magnetism. This peculiar property of some rocks was noted by Pliny the Elder (AD 23–79) in his *Naturalis Historia* – at thirty-seven volumes one of the largest single works to have survived from Roman times – he said his scope was 'the natural world, or life'. In about 1000 BC, Pliny says, a Cretan shepherd called Magnes was walking on Mt Ida in Mysia (now Turkey). Suddenly he was drawn to the Earth by the tacks in his sandals. He dug into the ground and discovered magnetite – later called lodestone – a magnetic oxide of iron. This was later to be called by some the Magnesia stone – the stone that points the way.

A magnet was a miracle that had connections with an inner world. Thales of Miletus believed that a magnet possesses a soul but the Romans sensed that lodestones could be put to practical use. Titus Lucretius Carus (*c.*99–*c.*55 BC), better known as Lucretius, has only one surviving poem. It is called *De rerum natura* (On the Nature of Things). He asks why some materials are immune from the lodestone's effects. 'Some are held fast by their weight

. . . [as in the case of] gold. Others cannot be moved anywhere, because their texture, due to the large air spaces between the atoms that form them, allows the effluence, emanation of atoms from the lodestone, to pass through intact.' Ptolemy, the Greco-Egyptian astronomer from Alexandria in the second century AD, writes in his *Geography*: 'There are reported to be as well ten more islands in a row, called the Manolial [perhaps Sumatra or Java] amid which they say that boats having iron nails are held in check because the Stone of Hercules is native there, and it is for this reason shipbuilding is done in trenches.' In *The Arabian Nights* it says, 'tomorrow we will come to an isle of black rocks called the Magnetic Mountain . . . and all the thousands of nails in our ten ships were suddenly wrenched away and flew to join themselves with the mountain. The ship opened out and fell asunder, and we were all thrown into the sea.' In AD 200 the priests of Samothrace were selling magnetic rings as a cure for arthritis. Saint Augustine (354–430) wrote in his book *The City of God*, 'the magnet, which by its mysterious and sensible suction attracts the iron . . . has no effect on straw . . . if such common phenomena be inexplicable . . . why should it be demanded of man that he explain miracles by human reason?'

The architect to Alexander the Great, Deinokrates, had a dream about using magnetism in a temple he was designing for Ptolemy II of Egypt. It was to be called the Arsinoeon. He dreamed of an iron statue that could be levitated by means of lodestones but died before he could start its construction. Deinokrates was nothing if not ambitious; his plan to transform Mt Athos in the Chalcidice Peninsula into a statue of Alexander 2,000 m. high holding a goblet from which the mountain streams would flow was halted by Alexander's own death. Cedrenus says that an ancient image in the Serapium at Alexandria was 'suspended by magnetic force'. Cassiodorus stated, 'in the temple of Diana hung an iron Cupid without being held by any band'.

Some archaeologists believe that the Olmec, the first major civilisation in Mexico, known for its colossal head statues, may have used magnets. A lodestone artefact found in Mesoamerica, radiocarbon-dated to about 1400–1000 BC, means they might have used it as a crude compass or for geomancy, a method of divination, which, if proved true, predates the Chinese use of magnetism for feng shui by a thousand years. It is part of a polished hematite (lodestone) bar with a groove at one end (possibly used for sighting). Today it points 35 degrees west of north, but may have pointed north–south when in one piece. Other hematite or magnetite artefacts have been found at pre-Columbian archaeological sites in Mexico and Guatemala. But it is in China that the compass was born.

Chinese literature mentions magnetism in the fourth-century BC writings of Wang Xu: 'The lodestone attracts iron', and they called a crude compass the 'chariot of the south'. It was written that the people of Zheng always knew their position by means of a 'south-pointer'. The first mention of a spoon, possibly a lodestone, observed pointing in a cardinal direction, is a Chinese work written between AD 70 and 80: 'But when the south-pointing spoon is thrown upon the ground, it comes to rest pointing at the south.'

The first clear account of magnetic declination – the fact that a compass does not always point true north (Halley's variation of the compass) – is in *Mr Kuan's Geomantic Instructor* of AD 880. Another text, the *Blue Bag Sea Angle Manual*, also has a description of magnetic declination. The earliest reference to a specific magnetic direction finder for land navigation is in a Song Dynasty book dated to 1040–44. There is a description of an iron 'south-pointing fish' floating in a bowl of water, a means of orientation 'in the obscurity of the night'. The *Wujing Zongyao* (meaning Collection of the Most Important Military Techniques) says: 'When troops encountered gloomy weather or dark nights, and the directions of space could not be distinguished ... they made use of the south-pointing carriage, or the south-pointing fish.'

In Shen Kuo's *Dream Pool Essays* (1088), geomancers are mentioned who magnetised a needle by rubbing its tip with lodestone, and hung the magnetic needle with one single strand of silk attached to the centre of the needle. Shen Kuo pointed out that a needle prepared this way sometimes pointed south, sometimes north. The earliest explicit recorded use of a magnetic compass for maritime navigation is found in Zhu Yu's book *Pingchow Table Talks* (1111–17): 'The ships' pilots are acquainted with the configuration of the coasts; at night they steer by the stars, and in the daytime by the sun. In dark weather they look at the south-pointing needle.' The earliest reference to a compass in the Middle East is attributed to the Persians, who describe an iron fish-like compass in a book from 1232. The earliest Arabic reference to a compass – a magnetic needle in a bowl of water – comes from the Yemeni sultan and astronomer Al-Ashraf in 1282 He also appears to be the first to make use of the compass for astronomical purposes.

The first European mention of a magnetised needle and its use by sailors occurs in Alexander Neckam's *De naturis rerum* (On the Natures of Things), written in 1190. In 1269 Petrus Peregrinus of Maricourt, a soldier in the army of Charles of Anjou, the King of Sicily, described a floating compass for astronomical purposes in his *Epistola de magnete*. He seems to have been the first to make a magnet in the shape of a sphere and examine its effect on magnetised needles as he moved them around it.

From ancient times it had been the habit of mariners to avoid sailing between October and April, due to the absence of clear skies during the Mediterranean winter. The longer sailing season resulted in a gradual but sustained increase in shipping and trade. By about 1290 the sailing season could start in late January or February, and end in December. The extra few months were of great economic importance, enabling Venetian convoys to make two round trips a year to the Levant instead of one. In 1300, another Arabic treatise written by the Egyptian astronomer and muezzin Ibn Sim'ūn describes a compass for use in finding the direction to Mecca. Arab navigators started using a thirty-two-point compass rose during this time. Magnetite occurs all over the world and there are large deposits in Scandinavia. The Vikings also used a compass but they kept its existence secret. There were rumours among medieval sailors that magnetite attracted ships and dissolved them.

For some, magnets held special healing powers. In the thirteenth century Bartholomew the Englishman (*c.*1203–72), author of the book *On the Properties of Things*, said that 'This kind of stone restores husbands to wives and increases elegance and charm in speech. Moreover, along with honey, it cures dropsy, spleen, fox mange, and burns ... when placed on the head of a chaste woman causes its poison to surround her but if she is an adulteress she will instantly remove herself from bed for fear of an apparition.'

But the behaviour of a compass was not always straightforward. Sometimes it did not point true north, its magnetic declination. As we have seen, the Chinese knew of the effect in the twelfth century. Navigating by the pole star, they noticed the compass pointed either side of its direction in different parts of the globe. A description of it was published in 1514 by the Portuguese explorer João de Lisboa in the '*Trato de Agulha de Marear*', contained within his *Livro de Marinharia* ('Treatise on the Nautical Needle', *Book of Seamanship*). He is believed to have been on Ferdinand Magellan's voyage around the world. The first person to lay down practical methods of determining the magnetic declination was Francisco Falero, or Faleiro, a Portuguese in the service of the Spanish navy, in the *Tratado del esphera y del arte de marear; con el regimiento de las alturas; con algunas reglas nuevamente escritas muy necessarias* (*Treatise on the sphere and the art of navigation with manual of altitudes with some very necessary written rules*). In the eighth chapter magnetic declination is discussed in detail for the first time in print and three methods given for its calculation. Magellan seems to have taken a manuscript copy of this book with him on his voyage around the world in 1519.

In 1538–41 on a voyage to the East Indies and the Red Sea a Spanish navigator named de Castro made forty-three observations of declination. He wrote: 'This science of navigation is poorly distributed among the men, or because they act like idiots, which for a long time and continuous exercise they reach many particulars, though with all their works are never to gain authority in their office, or those who have no experience, but with much learning and great practice in the science of mathematics, reached the shadow of this art but not the true science.' The magnetic declination for Rome was recorded in 1544 by Georg Hartman in a letter that notes another curious property of the compass needle: it does not stay in a horizontal direction, it often points down, something called its magnetic dip. In 1581 Robert Norman in London published *The Newe Attractive*, a small book that was the first printed work purely on magnetism. He mentions that magnetic dip was discovered in 1576.

21

The Terrella

As private physician to Queen Elizabeth I of England during her final years, life for William Gilbert must have been difficult. He was the most respected physician in the land and was ideally suited to the post, having attended the final illness of Lady Cecil in 1589 as well as her influential husband, Robert Cecil, a decade later. In 1599 he became president of the Royal College of Physicians and was consulted far and wide over medical matters. In 1601 he became Elizabeth's doctor. The Queen was in reasonable health until the autumn of 1602, when she was sixty-eight years old, but in the following months she fell into a 'settled and unremovable melancholy', as well as alternating 'between fits of rage and periods of silence and stupor'. Gilbert was often in the firing line of Good Queen Bess's temper and he sought refuge in his library, for he was interested in many other things as well as medicine. One of them was magnetism. Some believe his interest in physical science was stimulated by meetings with Giordano Bruno, the free-thinking priest who was eventually burned at the stake in Rome for heresy, and perhaps by discussions with the mathematician Thomas Harriot.

In his study he possessed a terrella, or 'little earth', a magnetic ball. He devoted much of his time and energy to the study of magnetism, which he described as a mysterious, invisible force. He suspended magnetised needles on strings and moved them close to his magnetic globe, noticing how they moved and changed direction. Based on his experiments he wrote a magnificent treatise of magnetism, *De Magnete, Magneticisque Corporibus, et de Magno Magnete Tellure* (On the Magnet and Magnetic Bodies, and on the Great Magnet the Earth), published in 1600. The terrella, he said, gave off 'rays of magnetic virtue'. Gilbert was also interested in magnetic dip. He supposed it depended upon latitude and would enable sailors to determine their latitude when it was cloudy and they could not see the sun at midday or the pole star at night. He later wrote, '. . . things metallick were hidden and the knowledge of the stones unheeded'.

In the late 1590s Henry Briggs (1561–1630), a professor of geometry at Gresham College in London, had published a table of magnetic inclination with latitude for the Earth. It agreed well with the inclinations that Gilbert measured around his terrella. Gilbert deduced that the Earth's magnetic field is equivalent to that of a uniformly magnetised sphere, magnetised parallel to the axis of rotation. However, he was aware that declinations were not consistent with this model. Based on the declinations that were known at the time, he proposed that the continents were centres of attraction that made compass needles deviate. He even demonstrated this effect by gouging out some topography on his terrella and measuring the effect on declinations. A Jesuit monk, Niccolò Cabeo (1586–1650), showed that if the topography was on the correct scale for the Earth, the differences between the highs and lows would only be about one-tenth of a millimetre. Therefore, the continents could not affect the declination. Then it was discovered that the declination of the compass was not constant; the declination of London had changed.

Despite his fascination with magnets Gilbert dismissed their use in medicine as quackery, saying it was 'evil and deadly advice'. It was said that his library was a collection of wonders. Following his death (from the plague), only a few months after Elizabeth's, it was left to the Royal College of Physicians, but it was completely destroyed in 1666 during the Great Fire of London.

So it was that in 1693, Halley, together with Benjamin Middleton, petitioned the Royal Society for their support of a worldwide oceanic voyage to observe the magnetic declination. The Royal Society agreed to help by supplying a small vessel; Middleton would assist in the cost of the voyage, and the observations would be made by Halley. After years of delays, the Royal Navy took charge of the voyage, and in 1698 King William III commissioned Halley into the navy as captain of HMS *Paramour*.

Halley's interest in the Earth's magnetism began in his youth and continued until the end of his life. In 1683, he wrote 'A Theory of the Variation of the Magnetical Compass', describing the magnetic declinations in various parts of the world based on the observations of sea captains and explorers. Halley gives a sample of fifty-five observations from forty-seven locations and discusses the direction and the rate of change of the variation of the compass. Halley knew the magnetic declination changed with time and included five readings from London spanning more than a hundred years.

Many people had a high opinion of Halley, observing he is 'of a middle stature, inclining to tallness, of a thin habit of body, and a fair complection, and always spoke as well as acted with an uncommon degree of sprightliness and vivacity'. Another wrote he 'possessed still more of the qualifications necessary to obtain him the love of his equals. In the first place he loved them; naturally of an ardent and glowing temper, he appeared animated in their presence with a generous warmth, which the pleasure alone of seeing them seemed to inspire; he was open and punctual in his dealings, candid in his judgment, uniform and blameless in his manners, sweet and affable, always ready to communicate, and disinterested', and he 'remained a young man all his life, in his drive, practicality, and enthusiasm'. Thomas Hearne's journal entry of 1728 says, 'Dr Halley (now in the 72nd year of his age) does not care to be thought old'.

On his voyage Halley recorded the latitude, longitude and magnetic declination during his voyages in two separate journals now in the British Library. There were four places in the Atlantic Ocean where Halley crossed the line of no variation, or the agonic line: near Bermuda, near the equator, west of St Helena and east of Tristan da Cunha. On 17 February 1700, he writes in his journal, 'I Determine the Latitud of the most Southerly of the Isles of Tristan da cunha 37°25'.' Halley also writes that there is no variation east of Tristan de Cunha. On 24 February 1700, he says, 'No variation 11 1/2 to the Eastwards of the Islands.'

Halley's 1701 map was a landmark in science – the first magnetic map of a wide region of the Earth showing magnetic declination. It was the first map printed and published with isolines, lines representing equal magnetic declinations. Because of the change of magnetic declination over time, revisions to the map were required after Halley's death. In 1745 and 1758, Mountaine and Dodson, Fellows of the Royal Society, undertook the task. To this day, Halley's Atlantic map is still used as a reference datum to study the changes in magnetic declination that have occurred in the past three hundred years.

Halley died believing that 'the whole globe of the earth is one great magnet having four magnetical poles, or points of attraction, near each pole of the equator two, and that, in those parts of the world which lye near adjacent to any one of these magnetical poles, the needle is governed thereby, the nearest pole always predominant over the more remote'. A portrait painted in 1736, when he was eighty, shows him holding his 1692 diagram of the Earth's nucleus and its shells. He died in 1742 aged eighty-five. His tombstone is in the Royal Observatory Greenwich, and it reads: 'Under this marble peacefully rests, with his beloved wife, Edmond Halley, L.L.D. unquestionably the greatest astronomer of his age. But to conceive an adequate knowledge of the excellencies of this great man, the reader must have recourse to his writings, in which almost all the sciences are in the most beautiful and perspicacious manner illustrated and improved.' But Halley does not lie by his tombstone. His unkempt and unmarked grave is in St Margaret's, the parish church of Lee, about thirty minutes' walk from Greenwich. Another Astronomer Royal, John Pond, was buried beside Halley in 1854, as was Nathaniel Bliss, the fourth Astronomer Royal.

Christopher Hansteen (1784–1873) was a Norwegian scientist who also mapped the Earth's magnetic field. In 1810, the Danish Royal Academic Society announced a competition on the following problem: 'Is it possible to explain the magnetic uniqueness of the Earth by one magnetic axis only, or is one forced to suppose several?' Hansteen delivered a thesis in which he assumed two axes and won the prize. To enable him to write his thesis, he extracted as much data as possible on the magnetic declination from old expedition logs and papers from as far back as 1600. He also included an atlas of different maps showing the variability of both the declination and the inclination between 1600 and 1800. He complained in 1819, 'The mathematicians of Europe since the times of Kepler and Newton have all turned their eyes to the heavens, to follow the planets in their finest motions and mutual perturbations: it is now to be wished that for a time they would turn their gaze downwards towards the earth's centre, where also there are marvels to be seen; the earth speaks of its internal movements through the silent voice of the magnetic needle.' So it is that measurements of the magnetic field on the surface made for the first time three hundred years ago show that our magnetic field is

changing, and now we have permanent magnetic observatories stationed all over the globe as well as data from
satellites.

In the nineteenth century our understanding of magnetism changed utterly. In 1820 Hans Christian Oersted
(1777–1851) was giving a lecture when he demonstrated
that electric currents produce magnetism by placing a
compass next to a wire that conducted electricity. Initially
he was not impressed, and neither were most of his audience; it seemed such a little effect. Later Oersted was to
write that he could not explain, given the subsequent importance of his discovery, why it took him three months to
follow it up. But when he did he carried out some of the
most important experiments ever performed in science.
On 21 June 1820 he wrote a four-page report and circu-
lated it to a few journals. Scientists were astounded. He
had shown that electric currents could generate magnetic
fields. It was then up to André-Marie Ampère (1775–1836)
to take it one stage further and suggest that there were
electric currents within the Earth that generated its magnetic field. It was James Clerk Maxwell (1831–79) who
established beyond doubt the relationship between electricity and magnetism and deduced a series of deceptively
simple equations that are the basis of electromagnetic
theory today. But what was happening inside the Earth to
produce its far-reaching magnetic influence?

Professor Jon Aurnou of the University of California at
Los Angeles became interested in geological science when,
as a teenager, he went camping in the Adirondack Mountains in upstate New York. As well as being beautiful, the

Adirondacks are a linear mountain range that are formed along tectonic plate boundaries. Aurnou took geology classes but found himself asking questions relating to geophysics. So he switched to physics. When he decided to consider a career in research he went to Peter Olson's lab at Johns Hopkins University; Olson was setting up an experiment to study the basic fluid dynamics taking place deep inside the Earth. He told Aurnou: if you come here, you will build this device and do the experiments. Aurnou did just that. I asked him how he imagined the molten iron in the mantle behaves. 'I think of a huge turbulent swirling mass of liquid metal, and yet the flow is probably very well organised. It's convecting strongly in a rapidly rotating system with a strong magnetic field. As it swirls around it amazingly generates and organises a magnetic field. It's incredible. Out of this turbulence comes a large-scale planetary magnetic field. It's an organised turbulence . . . and honestly I don't know what that means.' Scientists have struggled for over a hundred years to understand just how the outer core makes a magnetic field.

In Belfast on a late November day in 2009, a bitingly cold but bright morning in a traditionally rainy month, a small crowd gathered on the Lower Antrim Road to unveil a plaque to a local mathematician and physicist. Present were scientists as well as relatives of Sir Joseph Larmor. They had a stone memorial plaque that had been rescued from the gatepost at Larmor's place of birth, a house in Adela Street, now demolished. They unveiled a new blue plaque stating that Larmor had lived in a building that had once stood on the site, and then they repaired to the elegance of the Diocesan Library in nearby St Malachy's College. In the speeches during the unveiling of the plaque it was mentioned that most patients in the hospital just around the corner on that morning's list for an MRI scan would have been unaware that the physics underpinning the process was first formulated by the man who had grown up in the house, to which Hugh and Hannah Larmor brought their young family in 1863 or 1864 from the farm at Ballycarrickmaddy. It was from here that their eldest child, Joseph, first went to the National School in Eglinton Street and from there, in 1869, to the Institute from where, years later, the fourteen-year-old matriculated at Queen's College, graduating BA in 1875.

At one time it was thought that the Earth was a permanent magnet, but in the 1830s Gauss analysed the structure of the Earth's magnetic field, describing it in powerful and elegant mathematical terms. It was clear from his formulation that the strength of the dominant field was changing and that it was not a permanent magnet. There are only two known ways to produce magnetic fields: by magnetisation and electric currents. The Earth cannot be a permanent magnet so it has to be generated by electrical currents deep inside it, and Oersted and Ampère had shown how electric currents might be responsible. Joseph Larmor (1857–1942) has a theorem, a formula, a frequency and a length all named after him and he was crucial in the understanding of what was happening in the Earth's outer core. In 1919 he showed that the Earth's magnetic field is produced by a kind of dynamo acting in its fluid iron core with symmetrical motions around an axis. He proposed that the same effect was responsible for the periodic eleven-year cycle of sunspots. But the idea ran into a problem. Thomas Cowling (1906–90) disagreed with the dynamo theory and produced calculations that showed Larmor was wrong. I saw Cowling give a lecture at Oxford University in the early 1980s and, even though elderly by then, he gave me the impression, correctly, that he still had more insight and ideas and was more productive than most of the audience. In 1934 he produced what has become known as the Cowling Anti-Dynamo Theorem. It said that symmetrical motion of currents couldn't generate a dynamo. There was no arguing with Cowling's maths, but it wasn't enough to kill the dynamo theory, as many thought it was too good an idea and that a way would be found around the anti-dynamo suggestion. Some argued that the motions of electrical currents in the Earth's outer core would not be symmetrical. They pointed out that the weather in the Earth's atmosphere wasn't symmetrical, so why would the weather in the core be?

Many regard German-born Walter Elasser (1904–91) as the founder of theoretical dynamo theory. In 1946–7 he published papers describing the first mathematical model for the dynamo. He conjectured that it could be a self-sustaining dynamo, powered by convection in the liquid outer core, and outlined a feedback mechanism between flows having two different geometries, basically north–south and east–west. But he could not generate a dipolar field. Cowling, it seemed, had won again. Elasser was at the University of Utah when he met a younger colleague, Eugene Parker (b. 1927), who took up the challenge. Parker is one of the premier astrophysicists of the twentieth century and beyond and his 1979 book, *Cosmical Magnetic Fields: Their Origin and Their Activity*, was almost always open on my desk when I was an astronomer. He got around Cowling's injunction by thinking of the outer core as not uniformly rotating. Elasser attacked the problem of the Earth's dynamo with rigorous maths, but Parker used physical insight. There was a problem with Parker's dynamo, however – it would grow out of control. It needed a feedback loop to stop it.

Hannes Alfven (1908–95) was a rebel whose rejections were better than most scientists' acceptances. He was blunt but not always correct; for instance, he didn't believe in the Big Bang. He calculated that if the Earth had been a permanent magnet it would have decayed in about 100,000 years, meaning that the magnetic field inside the Earth must be self-sustaining. He added a new insight to the way dynamos work with his 1942 idea of 'frozen-in' magnetic fields. When a magnetic field is immersed in a very good conductor of electricity, which can be a gas or a fluid, the field and fluid, he maintained, must move together. Any relative motion between them induces strong electrical currents that oppose the motion.

Edward Crisp Bullard (1907–80) made many contributions to the study of the Earth. He used seismology to study the sea floor, was one of the first to find evidence for continental drift and developed the theory of the dynamo. He had been working on the dynamo problem for several years when he became head of the UK's National Physical Laboratory in 1950. The NPL had been building what was the world's most advanced computer which they called the Automatic Computing Engine (ACE). It had been designed by Alan Turing. The first stage of the project was to build Pilot ACE, which was ready in 1950 and unveiled to the public in a series of press conferences when it was described as an 'electronic brain'. Bullard said that he would be glad to hear from anyone with an important problem to be solved which required lengthy and intricate arithmetical calculations. One of them was his own dynamo. Using Pilot ACE in 1954 he and Harvey Gellman produced the first crude though convincing dynamo.

Then, in 1955, Parker cracked it. He proposed that convection in the core moves in a similar way as hurricanes do in the atmosphere. As the Earth rotates, Coriolis forces cause hurricanes to rotate clockwise in the northern hemisphere and anticlockwise in the southern. He believed a similar thing happens in the core with the motions of the molten iron, and that magnetic fields would merge, cancel and leave behind a dipole field. It worked after a fashion, but it was too simple as it neglected turbulence. It was many years later that Steve Childress at New York University and Glen Roberts independently in 1970 proved dynamos can exist and are common in planets and stars. In the 1970s and 1980s many mathematical dynamo models emerged; they all have their merits in the way they treat how convection cells arise by the flow of heat from the interior, how they change, produce and interact with magnetic fields. But once the real world is let in, in the form of turbulence, things get a lot more complicated. There is a saying among physicists that if you come across turbulence do anything to avoid it! Consequently, no completely satisfactory dynamo model has been produced.

22

Reversal

Nowadays, except on very rare occasions, the aurorae – the northern or southern lights – are confined to a region around the magnetic poles because the solar particles that cause them are electrically charged and are attracted to the magnetic poles. But there will come a time when aurorae will be seen all over the world; as well as Reykjavik and Anchorage they will be frequent visitors to Nairobi and Singapore. In the past the Earth's magnetic field has weakened and 800,000 years ago compasses would have pointed south instead of north and some of the time compasses would have been of no use. One day our magnetic shield will not be as protective against harmful radiation from the Sun or from deep space. Then, slowly, after thousands of years, scientists will notice the Earth's magnetic field is starting to increase and over more thousands of years it will return but with the previous magnetic pole configuration. It is going to happen, but don't hold your breath.

The idea that the Earth's magnetic field will soon flip and cause havoc is a staple of many newspapers, and for many journalists. The headlines often say something about the collapsing magnetic field that could affect the climate and wipe out power grids. Scientists, they add, say a flip is overdue, and although they are unsure about when it might happen it could occur anytime, punching a hole in the ozone layer and driving up rates of cancer. Imagine, say experts, for a moment your electricity supply was knocked out for a few months – very little works without electricity these days. It's a story that appears often, perhaps after details are revealed about satellite measurements of the Earth's magnetic field, perhaps someone is writing a book, or perhaps a journalist is just looking around for a story. Whatever the motivation, it's now ingrained in the consciousness of many people that dire things are about to happen to the Earth's magnetic field; after all, it has happened before.

Data from magnetism in rocks and from ancient campfires tell us that the polarity of the Earth's magnetic field remains stable for long periods – called chrons – and then there is a reversal. Chrons can last for almost any length of time. In the geological record there is evidence for them lasting as long as a million years or as little as 100,000 years and there doesn't seem to be a pattern regarding their duration. The latest one is called the Brunhes–Matuyama reversal and it happened 781,000 years ago. It's named after Bernard Brunhes (1867–1910), a French geophysicist who discovered that the Earth's magnetic poles flip, and Motonori Matuyama (1884–1958), a Japanese geophysicist who realised it had happened many times in the past.

In 1906 Brunhes saw that, as they cool, newly baked

bricks contain iron-rich mineral particles that align themselves with the direction of the Earth's magnetic field. He was not the first to notice this. English physicist Robert Boyle may have been the first to recognise it in 1691. It seems that Brunhes had been very impressed with the work of Giuseppe Folgheraiter (1856–1913), who had shown that ordinary bricks and pottery carried a particularly strong and stable remnant magnetisation that is aligned with the direction of the magnetic field in which they were baked. Folgheraiter was especially interested in the magnetism he found in Etruscan, Greek and Roman vases, urns and amphoras. He believed they could be used as guides to the Earth's magnetic field in the past.

Earlier Brunhes had been examining rocks under an ancient lava flow at Cézens in the French Massif Central when he noticed they were magnetised in the opposite direction to the Earth's present-day magnetic field. He deduced that today's North magnetic pole must have been in the south when the rocks solidified and retained the ambient magnetic field direction at the time. He presented his ideas at a meeting of the Société Française de Physique, and the talk was published in the *Journal de Physique* in November of the same year under the title 'Recherches sur la direction de l'aimantation des roches volcaniques'. He also sampled several sites of naturally baked clay, at Beaumont, where the clays were baked by lavas of the Montjoli volcano, and near the village of Boissejour, where the clays are beneath the lavas of the Gravenoire volcano. His conclusion was that: 'at a certain moment of the Miocene epoch, in the neighbourhood of Saint-Flour, the North Pole was directed upward: it was the South Pole which was the closest to central France.' This is the first suggestion ever reported that the Earth's magnetic field had reversed itself in the geological past.

In 1904 the journal *Nature* announced the death of Brunhes: 'It is with great regret that we have to announce the death of M. Bernard Brunhes, the director of the observatory of the Puy de Dôme. M. Brunhes died at the early age of forty-seven . . . Under his directorship the observatory won a prominent position for researches in the several departments of terrestrial magnetism, the physics of the earth's crust, and the exploration of the upper atmosphere.' Despite this it took half a century before the scientific community accepted his ideas.

The explorer Paul-Louis Mercanton (1876–1963) realised that if the geomagnetic field had reversed in the past, then reversely magnetised rocks should be found all over the world. Between 1910 and 1932, he studied lavas from Spitsbergen, Greenland, Iceland, the Faroe Islands, Mull and Australia, finding both normal and reversed magnetisation in both hemispheres, showing that Brunhes' initial observations were correct and providing evidence that field reversals are a global phenomenon. Acknowledging Mercanton's earlier work, Matuyama said: 'According to Mercanton the earth's magnetic field was probably in a greatly different or nearly opposite state in the Permo-Carboniferous and Tertiary ages as compared to the present. From my results it seems as if the Earth's magnetic field in the present area has changed even to the opposite direction in comparatively shorter duration in Miocene and also Quaternary periods.'

We now know that the Earth's magnetic field has flipped hundreds of times in the past. The Brunhes and Matuyama periods have now been examined in more detail and in the Matuyama period there are further reversals that have now been recognised. In the last two and a half million years there have been eleven reversals. A statistical analysis of them shows that there is no pattern, they are random, a fact that tells us something about how they occur. With this in mind we can't predict when it will happen again, we can just look for the signs. According to some they are there.

Keeping track of the North magnetic pole isn't easy. As we shall see, compasses do not point due north, but some way off it towards the magnetic pole. The problem is that the magnetic pole is moving and it is important to track it, not just for users of compasses – still useful even in these days of satellite-based global positioning systems – but to understand what is going on thousands of kilometres beneath our feet where the Earth's magnetic field comes from. Every so often the Canadian government dispatches a scientist to find the North magnetic pole.

The last person to do this regularly was Larry Newitt. He used to set off on a seven-hour flight from his base in Ottawa to Resolute Bay, the closest inhabited spot to the North magnetic pole. Then he took a three-and-a-half-hour flight north in a Twin-Otter aircraft that can land on pack ice. The magnetic pole is currently at sea and can be reached only at the end of the winter when it is frozen.

Each time he went back it had moved. 'We're following it across the ice,' he told me at the time. 'It jumps around from day to day and year to year and we have to keep track of it.' The landing is tricky and he and his colleague put the aircraft down as close as they can to where they think the magnetic pole is. By placing magnetic sensors on the ice Larry was able to surround the magnetic pole and triangulate its correct position. In recent years, the field has been behaving in ways not previously seen in the short time it has been monitored in such detail. Measurements of the magnetic pole's position taken in 1904 by explorer Roald Amundsen put it in roughly the same place as an earlier though less accurate measurement made in 1831 by the British explorer John Ross. Since then it has meandered northward until about forty years ago, when it started behaving differently. 'There was a slow drift northward, but it then started to move faster. It is now moving northward, away from Canada to Siberia, at a rate some four times faster than it used to,' said Dr Newitt. If it is a normal reversal it will take a thousand to 10,000 years to reverse, but sometimes it can happen a lot faster.

About 41,400 years ago, give or take 2,000 years, during the last Ice Age, something happened to the Earth's magnetic field, something remarkable and very, very swift. Observations of the magnetic properties of sediments laid down in the Black Sea, and beryllium and carbon isotopes in Greenland ice cores, indicate that a geomagnetic reversal took place with astonishing rapidity. In just 250 years the Earth's magnetic field declined by 95 per cent, flipped for 440 years when the North magnetic pole was at the South and vice versa and then over another period of 250 years or so things reverted back. Scientists call it the Laschamp event after the Laschamp lava at Clermont-Ferrand in the Massif Central, and it's a mystery. Since the Earth's magnetic field has been declining since Halley's time, and with further evidence that it was much stronger in Roman times, could we be in store for a flip, either a fast or normal one?

Not according to Cathy Constable of the Institute of Geophysics and Planetary Physics, Scripps Institution of Oceanography, University of California, San Diego. She points out that the Earth's present-day magnetic field is much stronger than it has been over the past million years or so, in fact more than twice as strong. Looking at the Earth's magnetic field over the past 10,000 years shows it to have been weaker until about 2000 BC, when over less than a thousand years it greatly increased in strength, at which it has stayed ever since. True, it has declined in the past few hundred years, but that decline is consistent with the up and down variations seen over the past few thousand years. This suggests that the Earth's magnetic field is not doing anything dramatic at present, despite what the headlines say from time to time.

About fifteen years ago Gary Glatzmaier, a fluid dynamicist at Los Alamos National Laboratory, and colleagues Paul Roberts (UCLA) and Rob Coe (UCSC), produced a computer simulation that took over 2,000 hours to run. The resulting three-dimensional numerical simulation of the geodynamo, run on parallel supercomputers at the Pittsburgh Supercomputing Center and the Los Alamos

National Laboratory, now spans more than 300,000 simulated years. The magnetic field that is produced has a dipole-dominated structure that is very similar to the Earth's and a westward drift of the non-dipolar field at the surface that is essentially the same as the 0.2 degrees a year measured on the Earth. In addition, about 36,000 years into the simulation the magnetic field underwent a reversal of its dipole moment, over a period of a little more than a thousand years. The intensity of the magnetic dipole decreased by about a factor of ten during the reversal and recovered immediately after, similar to what is seen in the Earth's palaeomagnetic reversal record. According to Glatzmaier, 'Our solution shows how convection in the fluid outer core is continually trying to reverse the field but that the solid inner core inhibits magnetic reversals because the field in the inner core can only change on the much longer timescale of diffusion. Only once in many attempts is a reversal successful, which is probably the reason why the times between reversals of the Earth's field are long and randomly distributed.'

Some scientists have even suggested that plate tectonics control the geomagnetic reversal frequency. They believe that when the continents are distributed asymmetrically over the Earth's surface there is a higher incidence of reversals. Perhaps they suggest it reflects lopsidedness in the mantle and the way heat flows out of the core influencing the currents inside it. Some computer models suggest that the frequency of reversals is linked to convection in the lower mantle. Peter Olson at Johns Hopkins University investigated the turbulent motions in the outer core that might lead to the geodynamo reversing, paying particular attention to the amount of heat passing through the core–mantle boundary. Seismic images were used to find warmer and cooler parts of the boundary. They found that when the heat transfer is high the fluid motions in the outer core are more turbulent, causing the magnetic field to destabilise. Science is about observation, theory and experiment. But how do you carry out experiments concerning the Earth's liquid metal core?

Despite what some people say there is a great deal of important research being carried out on the International Space Station (ISS). Its main purpose, one could argue, is to teach us how to live and work in space, how to keep astronauts healthy and functional, safe and productive. How to repair, replace and extend their working environment from the inside and with space walks, and how to receive and dispatch cargo and waste. In a way the ISS has to do nothing else to justify its existence, but it would be a greatly wasted opportunity not to do science onboard. The ISS is never going to be a centrepiece for astronomy or Earth observation science – these are best done with unmanned satellites – but there is much to be done concerning the adaptation of the human body to weightlessness as well as basic biological mechanisms, and also studies of the Earth's molten core!

The ISS's Fluid Science Laboratory facility is one of the major experiment suites located in Europe's Columbus laboratory section. One of the experiments is called Geoflow. It's a representation of Earth (or other planet) in miniature, being silicone oil held between two concentric spheres, which rotate about a common axis. A high voltage is applied between the spheres, creating a force that plays the role of gravity. Holding the inner sphere at a higher temperature than the outside one creates a temperature gradient from inside to the outside, as happens on Earth. Understanding the flow of the silicone oil in such conditions and its geometry allows insights into the movement of the Earth's mantle and core. Results from Geoflow are also useful for engineering applications, such as gyroscopes and bearings, pumps and high-performance heat exchangers. After thirteen months of continuous operations GeoFlow and later GeoFlow-2 have produced some intriguing results, helping to improve computer models of the flow in the mantle and the outer core.

At the University of Maryland, Dan Lathrop and his students are building a 3-m. titanium sphere which they will fill with liquid sodium that, when spun, they hope will give birth to a magnetic field similar to that generated by the Earth. The idea is that the sodium – which melts at temperatures less than boiling water – mimics the behaviour of the liquid iron in the outer core. In a way, according to Lathrop, it's a brute-force way of solving the problem of how the dynamo is made. 'In the end equations you write down can't be solved. Computers are too small and not quick enough and then there is turbulence which we don't really know how to deal with and may never know,' he told me, 'so we're building this machine to see if we can make a dynamo work.

'We run our machine about one week a month. Last week we ran it and took readings. It takes a week or more to look at the data then we make improvements and then we go back and run some more.' Lathrop points out that the experiment would not work if you make the inner sphere hotter than the outer one, like the Geoflow experiment on the International Space Station, because that would cause buoyant convection in the Earth's gravity. The way they get around this is by driving the inner sphere at different rates to the outer one, and this sets up differential movement that is not influenced by gravity and is technically a different form of convection. Will the churning sodium generate a self-perpetuating magnetic field? The apparatus uses Earth's natural magnetism as a 'seed field' to kick-start the process. As it is dragged and stretched by the spinning, conducting sodium it will generate electric currents. If a dynamo is created those currents will then generate additional magnetic fields that, when sufficiently twisted around, can amplify themselves and drive the process forward. No one knows if it will work, says Lathrop, 'there are neither theory nor experiments at these parameters and conditions'. They have found that magnetic fields are amplified in the apparatus but so far there is no dynamo effect. Lathrop's gut feeling is that he is not far from seeing a self-generating dynamo in the liquid sodium. 'Dynamos are easy to generate in nature,' he says, 'the same is not true for the lab.'

The Maryland experiment expands on earlier ones by the École Normale Supérieure institutes in Paris and Lyon and the French Atomic Energy Commission in Saclay. Built in Cadarache, France, the experiment uses spinning iron discs to force turbulence in liquid sodium within a cylindrical geometry. In 2007, the researchers reported they had created a dynamo that had some similarities to the Earth. 'Our experiment is a physicist's experiment,' says Jean-François Pinton, a scientist working on the project. He says they only want to generate a dynamo, regardless of how realistic it is. Dan Lathrop adds that he 'is after what the Earth does'.

Jon Aurnou's group is also working on a convection-simulation experiment to investigate how real geodynamos get started. It's turbulent motion in the presence of a rotation and a magnetic field. That's a hugely complex problem, he told me. 'I try to study the simplest component we can take out of that.' His group has a cylinder filled with liquid gallium – this melts at 30 degrees C – which they heat from below and cool from above. 'So now it's convecting like a little parcel of planetary core. We rotate it and add a magnetic field. It's a simple model of the earth's outer core. We are not trying to make a dynamo, we impose a magnetic field and rotation and let convection occur. Dan Lathrop is investigating the magnetic field, we are investigating the fluid flow.'

The dynamo experimenters hope that their work will shed light on how rotational forces in the Earth's outer core deflect currents of liquid iron into a configuration that produces a magnetic field with North and South poles. It might also help to explain what triggers geomagnetic reversals.

Many believe we are seeing hints of structure in the outer core. Studying how the Earth's magnetic field changes can be traced back to changes in the currents of molten iron in the outer core. Several studies indicate that there is a regular variation in the geomagnetic field with a periodicity of sixty years, which is taken to be due to waves in the liquid iron. Calculations concerning the stability of those waves suggest that the upper regions of the outer core are layered. Others believe there could be large-scale cyclonic motions in the northern hemisphere.

The car park based at an altitude of a thousand metres is the place to start your circuit of Vesuvius's crater. There is a steep 1.5-km-long track to the crater rim that should take around thirty minutes to an hour and for which you will have to pay a fee. The walk round the AD 79 crater is spectacular as you look down into the chasm that is over 300 m. deep. Lavas from the 1944 eruption partially cover the crater floor and you may see fumaroles. If you wish to descend to the crater floor you have to contact the Club Alpino Italiano, which may provide a suitable guide. It was not always so easy to get a glimpse into the destroyer of Pompeii.

In 1638 a group of Jesuit priests made their way up the lava and cinder slopes of Mt Vesuvius carrying ropes and a large basket. Smoke was rising from the crater and as they climbed closer to its lip every one of them knew just how unpredictable this untrustworthy volcano was. With every step each man prayed, none more than thirty-six-year-old Athanasius Kircher, one of the priests. He was to be the passenger in the wicker basket who was to be lowered into the steaming crater as part of his investigations into what lay at the centre of the Earth.

For the past four years the astronomer Galileo had been under house arrest in Arcetri, near Florence, for heresy, his crime being to suggest that the Sun and not the Earth was the centre of the cosmos. For the first three years of his incarceration he was ordered to read the seven penitential psalms, until his daughter received permission to do it for him. Also during his incarceration he had written the book *Two New Sciences*, which was smuggled into Holland for publication to avoid papal censorship, but by the time the Jesuits made their weary way up Vesuvius Galileo had gone blind, probably because of medical neglect. Such was the fate of those who stood against the Church's doctrine on the way the universe was ordered. This was a thought uppermost in the mind of Kircher who, as a Jesuit, was an intellectual defender of the Church. Soon he would be lowered into the volcano and, according to some, closer to hell itself.

Athanasius Kircher (1602–80) was a politically influential German Jesuit, occultist and polymath working on the cusp of the Enlightenment. For some historians of science he is a curious footnote to the usual portrayal of the steady advancement of knowledge whereby observations brought new data about the real world that contradicted religious dogma and Aristotelian doctrine that had to be discarded in the face of scientific progress. It was not a straightforward or comfortable task, as Galileo knew all too well. In the seventeenth century and afterwards science was moving towards a realisation of the immensity of time and its relationship with the inner workings of our planet.

Kircher did not quite fit into this conventional view of scientific history. He published his first book, *Ars Magnesia*, about his research into the properties of magnets in 1631 and a few years later he was summoned to Vienna by the emperor to succeed Johannes Kepler as mathematician to the Habsburgs. However, the appointment was rescinded and he then worked at the Collegio Romano.

By early afternoon the makeshift ropes and crane had been fashioned and Kircher climbed into the basket with his notebook and Bible, said a prayer and gave the order for him to be lowered through the steam and noxious gases that occasionally wafted by. He made drawings and notes as he moved closer to the rumbling lava and the gates of hell. Looking down, he wondered how he was to explain what he saw.

Escaping the wrath of Vesuvius, a few days later he witnessed the volcanic eruption from fifty miles away that would certainly have killed him and his party. He likened the volcano's heat to that of the alchemist's furnace, its stench to the sulphur and bitumen fumes that he inhaled in his laboratory. He had it in mind to write the most important book yet about the interior of the Earth, but was the devil really down there lurking in the fires that stoked Vesuvius?

Kircher never quite made the leap into science, as his task 'was nothing less than to penetrate the workings of the Divine Mind'. His resultant work, his immense bestselling two-volume tome of 1664–5, *Mundus Subterraneus* (The Subterranean World), includes astrology, volcanoes, alchemy, dragons, eclipses, fossils and gravity. Yet he also includes observations throughout his work, which has been described as 'a textbook in general science which does not broach new frontiers of knowledge but proffers its information in a readable and lavishly illustrated form, free from mathematical and philosophical complexities'. Others have called it a reflection of 'what the seventeenth-century man in the street may have thought of the Earth's interior'. Curiously, much of *Mundus Subterraneus* has never been translated into English from the original Latin.

Mundus Subterraneus is a bridge between medieval thought and the growing empirical movement, which today we call the Scientific Revolution. As well as dragons and astrology, throughout the *Mundus* Kircher included descriptions of alchemic laboratories and experiments with rocks and chemicals that he said were analogs for an interpretation of the Earth. Hell had been invited into the laboratory. Later it would be said by the great Scottish geologist Charles Lyell that the present was the key to understanding the Earth's past. Before that Kircher had shown, for those who realised it, that rather than the present being 'the key to Earth's past', it was the modern laboratory that was the key to Earth's past. Then it was the rituals of alchemy, now it is the techniques of the diamond anvil cell.

Mundus Subterraneus contains some remarkable illustrations. Kircher's drawings of Vesuvius are spectacular, as are his cross-sections showing the interior of the Earth that he sees as a seething pool of magma with various channels to the volcanoes on the surface. He called his map of the centre of the Earth the 'Pyrophylaciorum', for the fire in the middle.

Kircher's examination of the Earth's interior rests on Plato's philosophy that the universe was fashioned by God the creator as a manifestation of his own perfection: 'by turning it he shaped it into a sphere . . . giving it the most perfect form of all.'

But the very centre of the Earth is not perfect.

23

Inside the Inside

'There is something here I can't begin to explain'

Professor Lidenbrock

In the novel by George Sand (the pen name of Amantine-Lucile-Aurore Dupin) published in 1884 called *Laura, Voyage dans le Cristal*, there are giant, unseen crystals in the interior of the Earth. The novel is a love story. Alexis, a mineralogist, is reintroduced to Laura, a cousin he has not seen since childhood, and he falls in love with her, re-writing the past and introducing the start of their love for each other as children. It is a work of fantasy, for Alexis and Laura enter a magical world inside a crystal, which is a place of truth. Nasias, Laura's father, appears and lures Alexis away to undertake a 'voyage of discovery and con-quest of the subterranean world'. Together they start a dreamlike journey and discover many beautiful places, but with the death of Nasias Alexis returns to the real world and marries Laura, but only after he has realised the true worth of real women and not his idealised vision of them. Curious, then, that there are giant and strange crystals at the centre of our world and that they were discovered by a woman struggling to be accepted in a male-dominated science.

A billion years ago no human, or indeed any advanced living creature, could have survived on our world. The 1,500-million-year reign of the Proterozoic era was drawing to its close as the Columbia supercontinent was breaking up and reforming into a new supercontinent called Rodinia – Russian for 'homeland' – whose rocks and strata geologists can still see all over the globe.

The earthlings back then were the archaeans, bacteria and the eukaryotes. They had a hard time surviving but by now they'd started to spread throughout the oceans, evolving, diversifying and becoming more adapted and resilient in many environments. Each of these domains of life was to split into many lineages, not all of which have survived to the present day. The divisions between plants, animals and fungi was established, and single-celled creatures which had lived a solitary, floating existence formed colonies and with them came a division of labour – cells on the outside of a colony evolved to take on different roles from those that lived in the centre. This was the template for the complex life form with specialised cells. In a way it was the start of us. Cells had also developed a new way to evolve by the direct swapping and mixing of genes – they had invented sex. Sponges lived in the seas and algal mats coloured blue and red cloaked much of the oceans and the shores of that long-dead continent.

Only a few forms of life had evolved to live on the land. At that time there was no ozone layer to protect those on the surface from harmful ultra-violet rays from the Sun. If you could walk across the barren starkness of Rodinia, over its multi-coloured sandstones and siltstones, there would be no life visible except for a few discolorations seen on the rocks, a form of fungus consisting of cells clumped into colonies the size of match heads.

At the end of the Proterozoic the Earth was on its third atmosphere. Life had added oxygen and although the atmosphere contained far less oxygen than it does today, it was in geological terms oxygen-rich. It spelled doom for the survivors of Earth's earlier times for whom oxygen was toxic, but it presented new opportunities for the others.

A billion years ago the Earth was a lot like Mars, which at that time also had a thick atmosphere, rivers, lakes, oceans and possibly primitive life like the Earth. But, as we shall see, what happened at the very centre of Mars took a slightly different course from that of our own planet, leaving it frozen and dead instead of warm and full of life. The different paths taken by these two worlds can be traced to events at their very hearts.

The Moon of a billion years ago would have looked different. Many of its more prominent craters had yet to be formed by giant impacts. It was much closer then and the tides it raised on the Earth were stronger. It took only twenty days to go around the Earth, not the twenty-seven it does now, and on the Earth each day lasted only eighteen hours.

As Rodinia was being assembled, as life entered the two-billionth year of its colonisation of our planet, a small event took place at the very centre of the Earth. It did so slowly, perhaps taking tens of thousands of years. The inner region below the mantle of the Earth – the core

– was molten iron but at the exact centre of our planet iron atoms started to clump together, initially atom by atom. It would have been barely detectable at first. A single crystal formed.

A billion years later, in the early months of 1933 to be precise, an American radio engineer built a contraption consisting of a radio receiver fixed to the focus of a steerable parabolic metallic dish. Pointing it at the sky he detected a faint hiss – radio waves from objects out in space. Radio astronomy, which has peered further out into space than anything else, was born. The Nobel Prize in Physics that year went to Erwin Schrödinger and Paul Dirac for their work on the quantum theory – the strange behaviour of matter on the tiniest scale. While waiting at a red traffic light in London one rainy day, physicist Leó Szilárd had the idea of an atomic chain reaction which led to the development of the atom bomb, and that year the most famous scientist in the world, Albert Einstein, emigrated to the United States as Hitler became Chancellor of Germany.

But in Søbakkevej, a tiny village just north of Copenhagen, a prim forty-five-year old woman, neatly dressed with short, tidy hair, was sitting in the sunshine next to a small whitewashed cottage with a red-tiled roof. It overlooked a wood and a quiet lake, pears were ripening in the trees and there was a scent of lavender. If Inge Lehmann was worried about what was to come she put it to one side and sought comfort in her garden. Whenever she was away for any length of time she would always say she was worried about the roses. She was unmarried, perhaps because of the central passion that ruled her life – earthquakes. She sat on the lawn at a big table, filled with brown cardboard boxes, pencils and sheets of paper. In the boxes were cards holding information about a particular earthquake and the times it was recorded by monitoring stations scattered around the globe. Her notes and calculations, as exact and as meticulous as any accountant's, were held down by stones used as paperweights. In the dappled sunlight that shone on to her desk she looked at the details of the shock waves, now familiar with their patterns and periodicities. She had a compulsion for silence as she methodically collated and analysed; she had an idea what she was looking for. It had first come to her a few years earlier but now she was close to acquiring all the evidence she needed to prove her case. Slowly a pattern was emerging, but as it became stronger she found it wasn't what she had expected. Inge Lehmann did not know it at the time but she had found the crystal at the heart of our world – or what that first crystal had become.

Later in life, when she had become established as the doyenne of seismologists, she recounted how she became interested in earthquakes. 'I may have been fifteen or sixteen years old, when, on a Sunday morning, I was sitting at home together with my mother and sister, and the floor began to move under us. The hanging lamp swayed. It was very strange. My father came into the room. "It was an earthquake," he said. The centre had evidently been at a considerable distance, for the movements felt slow and not shaky. In spite of a great deal of effort, an accurate epicentre was never found. This was my only experience with an earthquake until I became a seismologist twenty years later.'

Inge followed the seismic waves that ripple around the Earth and delve deep into it after a major earthquake. She knew that, as a result of the presence of the Earth's liquid inner core (discovered by Richard Dixon Oldham and discussed in Chapter 19), there was a shadow region on the Earth's surface between 120 and 144 degrees from every earthquake. No seismic waves could reach that region but in the records on the card files she kept finding faint flickerings where none should have been found.

On Monday, 17 June 1929 the so-called Buller earthquake struck New Zealand. It was a strong quake felt all over New Zealand and it killed seventeen people, mostly as the result of landslides. The epicentre was located close to the small town of Murchison, near the north-west end of South Island, but people reported feeling the tremor over a wide area. Most of the population of 300 in Murchison lost their homes. Exactly one hour later seismometers in Copenhagen picked it up. Seismic waves were picked up at Swerdlowsk and Irkutsk stations in the USSR that were in the shadow zone. They showed up in more and more earthquakes from all regions around the world. 'I preferred to read phases from borrowed records or from copies of records that had been obtained. It meant a lot of work, but the published readings were not always satisfactory, especially when the movement was complex.' What was happening? How did the seismic waves get there?

Dr Inge Lehmann was born in 1888 by the lakes in Copenhagen. Her family had its roots in Bohemia and its

Danish branch produced many barristers, politicians and engineers. Her father, Alfred Lehmann, was a professor of psychology at the University of Copenhagen. He was a severe and aloof man, rarely seen except at mealtimes, though sometimes on Sunday he took the family for a walk.

Inge entered the University of Copenhagen in the autumn of 1907 with high family expectations. She passed the first part of the examination in 1910 and, in the autumn, was admitted to Newnham College, Cambridge, for a one-year stay to complete her studies. She enjoyed England, 'in spite of the severe restrictions inflicted on the conduct of young girls, restrictions completely foreign to a girl who had moved freely amongst boys and young men at home'. But she did not complete her work. Stress forced her to return home in December 1911. Her recovery was slow; friends later said she cared too much about her work and pushed herself too hard. For a long time afterwards she did not know if she would complete her studies.

Fate gave her the exact set of skills she would need to make the most extraordinary discovery about the Earth up to that time. She worked in an actuary's office and then returned to the University of Copenhagen, graduating in 1920. In February 1923, she became assistant to the professor of actuarial science. Then came the turning point. In 1925, she was appointed assistant to Professor N. E. Norlund, who, as Director of 'Gradmaalingen', was planning to have seismographic stations built near Copenhagen and at Ivigtut and Scoresbysund in Greenland. So Inge entered seismology. 'I began to do seismic work and had some extremely interesting years in which I and three young men who had never seen a seismograph before were active installing seismographs in Copenhagen and also helping to prepare the Greenland installations. I studied seismology at the same time unaided, but in the summer of 1927 I was sent abroad for three months. I spent one month with Professor Beno Gutenberg in Darmstadt.' Meeting Gutenberg was an inspiration for her. In 1914 he had used data from distant earthquakes to infer an average depth of 2,900 km to the boundary of the Earth's central core, just 50 km short of today's estimate.

She soon realised that most earthquake data was not very much use for detailed analysis. For example, it was often impossible to determine the earthquake's epicentre to any reasonable accuracy. To minimise the errors others were making, she compared visually similar waveforms from different seismograms. In this way she believed one experienced person's coherent view would overcome that of many sloppy individuals. Seismic traces from earthquakes are complicated. As well as the initial arrival of the shock waves there are later arrivals that have taken many different routes through the Earth, having been reflected from the various layers within it. In addition, the speed of the shock waves varied. In the next few years using her new method she noticed that more and more often she was finding faint seismic waves where she should not have found them.

Slowly a picture emerged. She suspected that the unexpected waves were being reflected from something deep within the Earth. She put down her suspicions in a letter dated 31 May 1932 to Harold Jeffreys at Cambridge University, probably the world's leading seismologist at the time. 'I suppose they could be explained by a discontinuity surface within the core,' she wrote, and later, 'that is a fact, though it might be an inconvenient fact.' Jeffreys was interested but not convinced. He mentioned it to a friend, a Jesuit scientist, who was likewise unimpressed. Jeffreys wrote back to her that he thought his friend would have 'jumped at the discovery of hell'.

By 1933 she had all the evidence she needed. She showed it to Gutenberg, by now ensconced at the University of California. He was convinced, as was his protégé Charles Richter. Jeffreys took a bit longer to be persuaded but by the time war broke out all agreed that Inge Lehmann had made a remarkable discovery – a new region of the Earth. The timings of the waves gave her an estimate of the distance to the reflection. It was deep, deeper than anything ever detected. But what was it? Was it solid?

In the years following the Second World War seismology made great strides, though not because of any heightened interest in the rocks and the workings of the Earth. When atom bombs detonate they shake the Earth, especially if they are underground. Following the development of the atom bomb, and especially after the USSR made one, the US realised that seismic monitoring stations could listen for nuclear tests as well as earthquakes, and were an essential tool for monitoring compliance with test ban treaties.

For decades after Lehmann's discovery little more had been found out about the inner core. It was small, and of the earthquakes strong enough to send shock waves deep enough to reach it, very little of their energy ever returned to the Earth's surface. Indeed, it was thirty-five years before proof was obtained that this inner core was actually solid, and that came from the Alaskan earthquake of 1964, one of the most powerful ever recorded.

The last thing written on the blackboard in the schoolroom on Chenega Island was 'Today is Friday March 27, 1964'. Every year the Pacific plate moves about 5 cm north-west below the North American plate, and the tension builds, compressing and warping southern Alaska, and every few tens or hundreds of years it snaps. It did this on that spring day in 1964, as the children of Chenega discovered.

It lasted just three minutes and was estimated to be a 9.2 magnitude quake, the second-most powerful ever recorded. Although there were only 143 deaths, many of them were children, especially those from the native settlement at Chenega, which was struck by the resulting tsunami, killing twenty-six of its sixty-eight inhabitants. The community lost its heart and moved to a new site more resilient to earthquake damage and tsunami. The old village still stands, what's left of it. The schoolroom is still there, or, rather, its shell, the date still on the blackboard.

The origin of the earthquake was some 12 km below Prince William Sound and it not only radically altered the coastal landscape but it also changed the direction of science, converting plate tectonics from a theory into a harsh fact.

The quake was detected by the new World Wide Standard Seismographic Network that had set its instruments to record long-period vibrations, providing an unprecedented view into the mechanics of a large earthquake. As we have seen, it was the aftermath of this earthquake that interested two seismologists, Adam Dziewonski, at the University of Texas, and Indiana-born James Freeman Gilbert. They wanted to study the earthquake but in those pre-digital days faced many logistical problems. The main one was that the data from each seismic station was recorded on photographic paper. This would have been no problem at all for Inge Lehmann, who got to know every seismic waveform from the stations that recorded the earthquake she was studying, but it was a big problem for Dziewonski and Gilbert.

When the initial burst of seismic waves had died down they noticed that the whole planet was left vibrating for days afterwards. It was as if the Alaskan earthquake had acted like the clapper of a bell that set the Earth ringing every 247 seconds. With an idea of the Earth's structure, its crust, mantle and core and their general properties the two seismologists could predict the timescale at which the entire planet would ring after a major earthquake. Try as they might they could not get an answer of 247 seconds. It was incompatible with the Earth having a liquid inner core. The only explanation was that Inge Lehmann's inner core was made of solid iron.

Inge Lehmann died in 1993 aged 104, withdrawn from the world and always embarrassed and unwelcoming of the attention paid to her. She carried on working despite increasing frailty. Her friend Erik Hjortenberg helped her write her last scientific paper in 1987, when she was ninety-nine years old. She participated in a reception held at the Danish Geodetic Institute on her 100th birthday. Geophysicists from all over the world attended.

24

Stranger in a Strange Land

'Almost everything known or inferred about the inner core from seismology or indirect inference is controversial.'

Don Anderson

Dive through the molten iron of the outer core, through its currents that are glacial to short-lived humans but swift on geological timescales, and eventually you will reach a solid sphere – an iron-clad world held within a metal sea and not obviously tied to anything above. Imagine the volume of water in all of the world's oceans and multiply it by five and you have the volume of the inner core of the Earth. No part of our planet has attracted so much scientific attention in the past two decades as the inner core. The story of its investigation is one we have seen repeated often in our contemplation of our planet. It starts off as simple – in this case a ball of solid iron – but as time goes on further observations reveal it to be far from straightforward. I have spoken to many scientists about this strange structure at the very heart of our planet; they have all said they can describe certain aspects of this 'ball of iron', but as to what it is really like, and what is really going on down there, they have little idea. In recent years the inner core has delivered to us a series of big surprises, and after all of them its status has risen. Small it may be – a radius of 1,220 km, and a surface area equal to that of Antarctica, which means less than 1 per cent of the volume of our planet and only about 2 per cent of its mass – but it has an importance out of all proportion to its parameters, and it packs more strangeness and mystery into its volume than any other region of the Earth.

It is difficult to gather information about the inner core. Although it's almost the size of the Moon it presents a small target for seismologists. Few seismic waves reach it, let alone return to the surface. New techniques are being developed to pick out the faint and elusive waves that reach the inner core and bounce back. Seismologists have borrowed ideas from the astronomers who regularly link telescopes to isolate faint signals in the sky. Likewise it is possible to correlate what seems to be background noise picked up by two or more seismic arrays. With the right processing it's possible to remove from the static faint echoes from the inner core, and the more we discover about it the stranger it becomes.

The lesson of the core is that, just as it becomes in any way familiar and perhaps comprehensible, it reveals a secret that draws another veil of mystery over it. The first surprise came in the early 1980s, and it was a very big surprise. After looking at the seismic wave travel time through the core of over 400 earthquakes spanning five years, scientists noticed something that shocked them. They didn't believe it at first and checked and rechecked the observations. Then they published the data but did not fully realise their significance. They found that there was a difference in the time it took seismic waves to travel through the core depending on the route they took: an equatorial route, say, from China to Central America, or a polar route, from, say, the South Sandwich Islands to Alaska. It wasn't by much, a few per cent or about five seconds, but it was significant. It was more difficult for seismic waves to travel east–west through the core than north–south, and initially scientists had no explanation for it. The solution to this puzzle radically changed the way we viewed the inner core. 'It was a very big surprise,' says Karen Lythgoe of the University of Cambridge, 'but the fact that two groups found the same thing using different techniques made the scientific community take notice.' Both groups of researchers who detected the effect came from Harvard University, signalling quite a scientific coup for that institution.

What might be an explanation came from interpreting the seismological data alongside other branches of science. Some speculated that the atomic structure of iron under Earth-core pressures might be the cause: specifically, the way the iron atoms are strung together in a lattice might produce a 'fast axis' when it came to allowing sound waves to travel through it. If the material at the core was aligned so that this fast axis was pointed roughly north south then it might be why seismic waves travel faster in that direction.

By the early 1990s supercomputers were capable of simulating iron under the conditions found at the core. During this time the number one computer of choice for scientists was the $30 million Cray C90 which was used for almost all big computing projects, from weather fore-casting to collecting data from the sub-atomic particle smashers at CERN. Four times faster than previous machines, it was capable of operating at gigaflop speeds (a gigaflop is a billion procedures a second, fast for the time but now somewhat slower than the computer I am currently using to type this). One of the most active C90s of the time was to be found at the University of Pittsburgh Supercomputing Center where Ronald Cohen and Lars Stixrude had a project to simulate the centre of the Earth. To carry out calculations to see how atoms of iron would behave when forced together under great pressure and temperature is not easy. There are dozens of things that need to be taken into account, especially quantum effects, which are often strange and seemingly illogical. 'To do these esoteric calculations,' said Stixrude, 'solutions which you can get only with a supercomputer, and get results you can compare directly with messy observations of nature and help explain them – this has been very exciting.' It's an approach emphasised by Maurizio Mattesini from the Complutense University of Madrid, Spain. 'There are two ways we can study the inner core,' he says, 'you can use seismology or quantum mechanical calculations to work out from first principles what the arrangement of iron atoms should be under such incredible pressure and then calculate how they let seismic waves pass through them.'

Hexagonal crystals are the iron atom's way of resisting extreme pressures when they arrange themselves into a rigid lattice whose inter-atomic forces counter the pressure. A few atoms arranged this way look symmetrical but as the lattice grows larger it turns into a crystal with a preferred direction. If these crystals had been oriented in random directions there would have been no difference between north–south and east–west seismic travel times. But if the crystals were aligned the effect could be explained. 'Hexagonal crystals have a unique directionality,' says Stixrude, 'which must be aligned and oriented with Earth's spin axis for every crystal in the inner core.' Lythgoe agrees: 'Crystals are anisotropic – we believe there are crystals, aligned or not aligned, but,' she cautions, 'it's not as simple as that.' It seemed a reasonable explanation even when a more detailed analysis of the seismic travel times through the Earth showed that the inner core's fast axis was not aligned with the rotational axis of the Earth but tilted some 45 degrees. Perhaps the inner core was not aligned with the larger Earth?

Then things became even stranger.

In 1997 it was discovered that the inner core had even more structure than its known directional preferences. It seemed that polar paths on one side of the inner core travelled faster than on the other side, something that some scientists now interpret as a wedge-shaped region of different material, probably iron with a different amount of other elements within it, or iron crystals of a different shape or size. Things were becoming more and more complex. 'Honestly, I don't know what is happening,' says Lythgoe, 'there are lots of individual ideas that explain some things but none that explains everything.'

In 2002 Maiki Ishii and Adam Dziewonski published observations that seemed to show there was yet another structure on the way to the very heart of the Earth. They called this newly discovered region right at the heart of our planet the innermost inner core. It had a radius of about 300 km and its fast direction was at 45 degrees to the axis. It caught people's imagination as they asked if this was the last major boundary to be found within the Earth.

Xiaodong Song of the University of Illinois also believes the Earth has an innermost inner core. To get at it Song needed as much data as possible from as many earthquakes as possible. 'To constrain the shape of the inner core, we needed a uniform distribution of seismic waves traveling in all directions through the core,' he said. Data from the earthquakes was added to a computer model, each one providing a transept of the Earth. 'This is the first time we have been able to piece everything together to create a three-dimensional view,' says Song. He maintains that the data show a distinct change in the inner core, clearly marking the presence of an inner inner core with a radius of 590 km, slightly less than half the size of the inner core. He believes it is there because the strain on the giant crystals becomes too great and they break down and merge. 'The inner inner core may be composed of a different phase of crystalline iron or have a different pattern of alignment,' Song said. 'For many years, we have been like blind men touching different parts of an elephant,' he says. 'Now, for the first time, we have a sense of the entire elephant, and see what the inner core of Earth really looks like.'

Not everyone agrees, however, pointing out that the

models and observations were inconclusive and in many cases inconsistent, consequently the existence of the innermost inner core remains to be proved. In 2013 Karen Lythgoe and colleagues carried out an analysis of seismic wave directions through the Earth. 'To study the inner core we need earthquakes above mag 6 just to get enough energy down into the inner core,' she said. Hence she examined every earthquake above mag 6 since 1990 that was accurately located in terms of depth and time – about fifty earthquakes – and looked at what seismic stations on the other side of the earth had detected. She found that no innermost inner core was required. The changes seen on the surface of the inner core reached all the way down.

Seismologists are very adept at gleaning every detail of information from seismographs and looking for faint signals from reflections for the Earth's internal layers. A way to get information about the inner core boundary is to look for seismic waves that have reflected from its underside. Recently, using data from a deep earthquake that took place in Indonesia, scientists stacked the seismograms obtained from various stations on almost the opposite side of the world from the quake, the theory being that seismic waves that reach the inner core can be reflected as they try to leave it by the inner core's surface. If this happened the seismic waves should be detected very close to 180 degrees from the quake. They were first detected very faintly in the 1970s. The problem is that there are only three locations that are opposite the epicentres of recurring deep earthquakes: northern South America, eastern central China and northern Africa. This didn't stop a team of Chinese scientists looking for them. Their analysis suggests deep layering within the inner core, but again not everyone interprets their data the same way.

The geography of this strange tiny world-within-a-world predominantly made of iron with a small amount of lighter elements is clearly complex. It does not look the same from all sides, indicating that there are major differences between its east and west sections. Clearly this is no simple ball of iron and its lopsidedness probably represents some combination of the way it was formed in these extreme conditions and how it has grown in response to the forces acting upon it.

A few years ago some of these strange properties were explained by questioning what was happening to the inner core, in particular how it got rid of its heat. It was concluded that it wanted to churn or convect like the rocks of the mantle but its high rigidity prevented it from doing that. The way it adjusted itself was by churning in a different manner by having one side melt and the other solidify, as if it was always trying to move sideways. Scientists called this translational convection. Some believe this behaviour forms a basis for explaining the east–west asymmetry. The present translation rate is found to be typically 100 million years for the inner core to be entirely renewed, and the process makes the exact centre of the Earth off by about a hundred metres.

Translational convection may have an effect on the magnetic field generated by the fluid motions of the outer core. Recent simulations of the process suggest that it makes the magnetic field in the Earth's central region asymmetric because the magnetic field will be stronger on the cooling side of the inner core. This in turn will mean that convection at the base of the outer core will also be lopsided.

In 2012 things changed again when a team from the University of Leeds, led by David Gibbons, and one from the University of London determined the properties of iron atoms assembled into crystal structures and they arrived at a far-reaching conclusion. The thermal conductivity – the ability to transport heat – of the inner core was about three times higher than had been thought. It might not seem a big change but it made a big difference to our understanding of how the inner core behaves and it required some of our most fundamental assumptions about it to be revised. It meant that the heat from the inner core was being lost by conduction and not convection, and therefore the inner core could not have been in translational convection mode. For a while it seemed that this new information had set back our understanding. But the following year it was demonstrated that you could have the convection driven by compositional variations in the iron – that is, by differences in the density of the liquid iron above the inner core due to it holding on to different amounts of lighter elements. Stranger and stranger.

One recent idea that might explain the differences between the two hemispheres of the inner core is down to Hrvoje Tkalčić, of the Australian National University, and Maurizio Mattesini. Tkalčić told me, 'I wanted to be an astronaut so I studied physics. Then I became fascinated by the inner core.'

Once again, they suggest it's down to the way iron

atoms pack together; specifically, that the inner core is made up of a mosaic of two ways that iron atoms pack together under such extreme pressure, in particular either hexagonal close-packed or the body-centred cubic iron crystals. 'The Candy Wrapper model,' as Tkalčić calls it, 'accounts for a dynamic picture of the inner core where different iron crystal shapes can be stabilised at the two hemispheres.' By comparing seismic data from over one thousand earthquakes across the globe with quantum mechanical models for the properties of iron, they suggest that seismic variations directly reflect variations in the iron structure. They propose that the eastern and western sides of the core differ in the mixing of these two kinds of structures. 'We showed that seismological data are best explained by a rather complicated, mosaic-like, structure of the inner core, where well separated patches of different iron crystals compose the anisotropic western hemispherical region, and a conglomerate of almost indistinguishable iron phases builds up the weakly anisotropic eastern side,' Tkalčić explained.

As if the inner core had not exhibited enough strange properties it might have added one more. There is some evidence that it does not rotate at the same rate as the rest of the Earth. On the one hand, you might ask: why should it? It is, after all, sitting in the middle of a sea of liquid metal and not physically connected to anything, so perhaps it could spin at a different speed, and with a different axis of rotation than the Earth around it. Over the past fifteen years or so some scientists have suggested this, based on observations of earthquake doublets. These are two almost identical earthquakes that occur in the same region many years apart. Being identical, they *should* produce identical seismograms, but sometimes they don't, and this is interpreted as the seismic waves having to bounce off a different region of the inner core's surface because it has rotated slightly in the period between the two quakes. Using this interpretation it is possible to put limits on the differential rotation of the inner core.

The inner core did seem to be rotating slightly differently from the rest of the planet above it. In the past fifteen years there have been several estimates ranging from one degree a year to between 0.12 and 0.38 degrees a year in an eastward direction. But the readings were not consistent; some earthquake doublets showed no differential rotation, others showed it rotating slightly westward. Also data from earthquakes in the Aleutian Islands observed in southern Africa show variations that are incompatible with anything else. Kuril events recorded in South Africa also do not show anything conclusive. Because of this some scientists believe that the inner core shuffles, and is sometimes faster and sometimes slower than the rest of the planet above it. A recent study published in the journal *Nature Geoscience* drew together data from twenty-four known double earthquakes as well as adding information from seven new ones. It indicated that while there was reasonable evidence that the inner core was rotating about a third of a degree a year out of step with the mantle above it, it also exhibited what they called shuffling, when it would spin faster or slower for ten years or so. One curious aspect of this latest work is that it indicates that the inner core's rotational behaviour in the last ten years has been very unusual in terms of it spinning quite a bit faster for a few years and then much slower, with the transition between the two speeds taking place in less than a year. To me this suggests that the data needs to be examined again.

But there are more surprises in store for us at the centre of the earth. The inner core, that ball of iron almost the size of the Moon, is a newcomer.

25

The Crystal Forest

So much complexity, so many baffling observations that cannot be reconciled into a single picture; the inner core recedes from us the closer we get to it. But if you gather up all the information we have it is possible to say something of its history, its influence and where it came from. And so this is its story.

After its formation the core of the Earth cooled slowly due to the insulating rock of the mantle above it. The Earth's heart, made of liquid iron gathered from the planetesimals and the great impactor three and a half billion years earlier, had been cooling at a rate of about 100 degrees C every billion years. It is a curious property of iron under such extreme conditions that increasing pressure reduces its melting point. So as the core cooled there came a time, a billion or so years ago, when, under a pressure of three million atmospheres and at temperatures as hot as the surface of the Sun, the conditions were right for iron to solidify at its centre. Thus the atoms of iron, used to coming together and being instantly torn apart by the high temperature, found that some of them stuck together and more iron atoms were added and so a single crystal, or perhaps more than one scattered over a short distance, began to form.

A billion years later, that crystal has spawned many others and has grown to a ball of the radius of 1,220 km. Today it is growing at a rate of about a millimetre a year by freezing about 5 million kg of iron out of the outer core and depositing it on its surface. That means that every second ten to the power of thirty-three atoms are leaving the molten outer core and sticking themselves to the inner core's iron surface; that is more, far, far more, iron atoms every second than there are stars in the entire universe.

Computer simulations suggest it could be growing at a greater rate at its equator than at its poles, so it could have an equatorial bulge. The crystallisation of the inner core is unusual because of the effect of pressure on solidification temperature. This means that the solid is actually hotter than the overlying liquid, so heat flows from solid to liquid, contrary to what is typical.

One of many experiments designed to get information about the inner core boundary was carried out recently by monitoring earthquakes in the Banda Sea near Indonesia. The quakes were recorded at the high-sensitivity seismographic network in Japan. After the disastrous Kobe earthquake in 1995, a high-sensitivity seismograph network was constructed, named the National Research Institute for Earth Science and Disaster Prevention. Hi-net uniformly covers the Japanese islands with a detector spacing of 20–30 km. In the Banda Sea seismic waves' observed variations in travel times of a few seconds were interpreted as being due to a height variation of about 14 km on the surface of the inner core.

Some believe that the boundary between the inner and the outer core is undulating and mushy, with the iron crystals growing dendritically and reaching out with a kind of iron mush between them. During the crystallisation process any impurities in the molten iron such as nickel atoms are excluded, and they accumulate in a slushy layer above the crystals whose presence can be detected by its effect on seismic waves. Some believe the topmost few hundred kilometres of the inner core consist of small crystals of iron, but the deeper you go the more times the iron crystals, which are magnetically attracted to each other, may merge, losing their individual identity as the giant crystals the size of cities are aligned with the Earth's overall magnetic field – roughly north–south. Some computer simulations suggest thin sheets of iron sliding off the equatorial highs and piling up against each other at higher latitudes, rather like mountains building on the Earth's surface where thin slabs of continental crust are stacked upon each other.

These crystals are marvels, unseen wonders of the solar system. If you could sail past them they might remind you of a geological structure similar to the basalt columns of the Giant's Causeway in Co. Antrim, Northern Island, though in some cases thousands of times longer and wider. 'There were papers published suggesting that the inner core was a single crystal and other papers suggesting that there is data about the length scale of the particles in the seismic data due to the scatter of seismic energy – crystal sizes range from few tens of metres to 20 km,' says Tkalčić. Single iron crystals the width of a city stretching the distance from London to Birmingham! At the other extreme these crystals are made of individual atoms of iron arranged in their latticework rows, billions of them alongside one another inside a single full stop. Inside the lattice there is a regular geometry stretching for tens or hundreds of kilometres. If you could shrink yourself to the size of the lattice it would seem as if you were lost in an infinite universe, stretching out in rows in all directions, and flexing and rippling as shock waves from earthquakes pass through them: everywhere the same, almost for ever.

A scientist involved in inner core research told me, 'We are just moving from one paradigm shift to another. The community is sometimes nervous in accepting that this world is more complicated than we think. Everything is getting increasingly complex as we get more data. The structure of the inner core – more complex. The anisotropy – more complex. The topography of the inner core – more complex. Its dynamics – more complex. We are seeing more and more complexity.'

But there is another problem. If the inner core is an essential component of the generation of the Earth's magnetic field then how did the Earth generate its magnetic field before the inner core came into existence a billion years ago?

There are many lines of evidence that point to the Earth's magnetic field being long-lived. Moon rocks of various ages contain isotopes that indicate if they were being partly protected by the Earth's magnetosphere. Samples brought back by Apollos 14 and 17 indicate that they were under such magnetic protection about 3.5 billion years ago, but not 3.9 billion years ago. A close analysis of the magnetic properties of the archaean rocks suggests they were formed in a magnetic field similar to that found today, complete with similar field strength and frequent reversals. How did a magnetic field generate before the inner core solidified?

One theory is that the titanic collision that produced the Moon may have had something to do with it. Much of the energy of the impact would have been converted into heat and it is likely that the Earth was completely melted. It might be that this initial shock provided the energy that the core has been using ever since and that such a collision might have been essential to start the Earth's dynamo. Some scientists speculate that, were it not for the store of that excess heat, the dynamo might never have started because there wouldn't have been enough energy for convection. Without the dynamo and its consequent magnetic field solar particles would have bombarded the Earth's atmosphere and stripped it away, as has happened to Mars, and allowed harmful radiation to reach the surface. With less energy plate tectonics might not have got started and little water would have been present at the surface. Without oceans the crust might have been too strong to fracture into tectonic plates, and the inside of the Earth would not have been able to cool in the way it has.

Does this mean that the formation of the Moon was essential to provide the energy for our planet to behave as it does, and develop into a place that was suitable for life? Are we here because of a series of very rare events and co-incidences? Scientists should be very wary of coincidences and heed the wise words of Agatha Christie, when she wrote in the novel *Nemesis* (1971), 'Any coincidence is worth noticing. You can throw it away later if it is only a coincidence.'

Perhaps a liquid-iron core doesn't need a solid inner core to produce a dynamo, but this leaves researchers asking: where does the energy for the liquid iron's con-vection come from? Other researchers think we might be looking in the wrong place. They suggest that a magma ocean might have formed on the early Earth in the middle of the mantle and moved slowly downwards to the core. Might the motions of currents in the magma ocean be enough to produce a dynamo effect? In addition, it is thought that a basal magma layer would suppress heat flow from the core and prevent a liquid-iron dynamo. As the magma ocean solidified it would have lost its dynamo action, but because it would have let more heat flow out of the molten core perhaps it then allowed the liquid-iron dynamo to start. According to this idea the Earth has had its magnetic field generated in three ways, first by the molten rock in the mantle, then in the entirely liquid core, and then with a solid inner core. This model suggests there may have been a pause in the generation of the geomag-netic field between 2.1 and 2.4 billion years ago, and there is actually some indication of this. If Earth had a magnetic field shortly after its birth it could have far-reaching im-plications. For example, if Earth had magnetic shielding from the Sun that early on, it may have had consequences for the development of life. The first living cells on Earth may have first appeared 3.5 billion years ago. Perhaps the origin of life was related to the stable and magnetically shielded surface environment.

The inner core and its influence of the heat traversing the core–mantle boundary might also hold the secret of the timing of the next magnetic reversal. It is believed that the flow of heat across the core–mantle boundary has an important role to play in magnetic reversals. Some believe that higher heat flow particularly around the equator causes more magnetic field variability and a higher rate of reversals. A reversal would not be completely unexpected even though such reversals are randomly distributed in recent geological times, though not over all geological periods. Simulations of the geodynamo suggest that the transition from periods of frequent polarity reversals, such as the Middle Jurassic, to periods of stability, such as occurred in the Middle Cretaceous, may have been triggered by a decrease in heat flowing across the core–mantle boundary. Some research suggests that the inner core stabilises the dynamo and reduces the frequency of reversals.

When I go to the cinema to see a science fiction film I, like many scientists, have to get into the spirit of the film and learn to forgive film-makers if they take liberties with the science in the pursuit of an entertaining plot and an enjoyable few hours' escapism. Having said that, there are some movies that a scientist just can't get along with and almost every few minutes provide a groan. Such a movie is *The Core*, about a voyage to the centre of the Earth.

In a poll of hundreds of scientists about bad sci-fi films *The Core* was voted the worst. Basically, the premise of the film is that the outer core of the Earth has stopped rotating and a team of scientists is dispatched in a revolutionary capsule to deliver nuclear weapons to the outer core to 'kick-start' it and get it generating a magnetic field again. The capsule has a hull made of 'Unobtainium', an indestructible metal that can withstand the heat and pressure of the core. I know. I know.

At the Earth's Core was written by Edgar Rice Burroughs fifty years after *Journey to the Centre of the Earth*. It first appeared as a four-part serial in *All-Story Weekly* in April 1914. In the story the author describes how, when travelling in the Sahara, he comes across a remarkable vehicle and its pilot, David Innes. The vehicle is an 'iron mole' and during its test run it is realised that the craft cannot be steered and it burrows 800 km into the Earth, emerging in the unknown world of Pellucidar perched on the outside of an internal hollow shell.

But could we ever construct a probe to reach inside the Earth, if not get to its very centre? A decade ago, Jet Propulsion Laboratory planetary scientist David Stevenson proposed a 'thought experiment' describing a 'grapefruit-sized' probe descending into the Earth's upper mantle. The idea was to use a crack in Earth's crust to pour a huge volume of liquid-iron alloy that would sink to the mantle under the weight of its own gravity taking the small probe along with it. Imagine a long, narrow crack filled with 100,000,000 kg of liquid iron – as much iron as is produced by all the refineries in the world in one hour. The weight of the iron would cause it to descend towards the Earth's core. Brief calculations indicate that the probe would get to the core in about a week. When it got there, the probe could measure the core's temperature, pressure and chemical composition. It would probably melt after less than a day there. It could communicate by creating tiny artificial earthquakes that could be detected from the surface. Intriguing, certainly, but just like in the movie it isn't going to happen. Forget the metal: a better way to design such an earth probe would be to make it hot so that it could melt its way down and not ride along in a stream of molten iron. The probe would melt its way through the Earth's layers in the same way a hot knife cuts through butter. Some scientists believe that it would have to be a rather large probe, about the width of an aircraft carrier, with a powerful plutonium reactor to heat the rock to a thousand degrees on a mission that would take about thirteen years to reach full depth. It would be a mission like no other; even designing a probe to withstand the harsh conditions on Venus's surface – 100 atmospheres at 500 degrees C – would be easier than designing one to go inside the Earth.

A design that is a little more practical is to construct smaller spheres packed full of radioactive Cobalt 60 that is readily available in spent, self-sealed radioactive sources used in medicine. Such a probe, launched from the bottom of a deep borehole, could bury itself by melting the rocks beneath it. If launched from the ocean floor it could conceivably get to a depth of about 100 km in a few decades. Designing a suite of sensors to measure temperature, pressure and composition while inside a sheath of molten rock would be problematic. Considering all the other proposals for science projects to study the Earth, I can't really see a probe into the Earth getting anywhere.

26

Other Worlds, Other Journeys

In many ways the journey to the centre of the Earth, travelling through the diversity of its crust, mantle, the outer and inner core, is the most interesting one in the solar system. There are similar journeys to be made from the surface to the cores of other planets. For some of them the journey is a variation on the one we have taken, but for others it is radically different, traversing stranger realms, encountering far greater extremes of pressure and temperature than are found within our own planet. Do we learn more about our world, or other worlds, by taking other journeys?

Astronomers divide the worlds of our solar system into four kinds. There are the terrestrial planets, Mercury, Venus, Earth and Mars. These are small rocky worlds nestling close to the Sun. Beyond the orbit of Mars lies the asteroid belt, populated by numerous irregularly shaped worlds – leftovers and remnants of worlds never formed or worlds torn asunder. Inside that belt lies Ceres, which at 950 km in diameter is regarded differently from the other bodies around it, and is classed as a dwarf planet. Further out is the realm of the gas giants,

Jupiter, Saturn, Uranus and Neptune, although some astronomers categorise the latter two as 'ice giants'. These are massive worlds, seriously giant planets, not primarily composed of rock but of hydrogen and helium with gaseous surfaces that thicken the deeper you go before you encounter inside them truly alien conditions. Beyond Neptune is the realm of icy, rocky worlds – more dwarf planets – that inhabit the solar system's cold, dark outer reaches.

The closest planet to the Sun is Mercury, tricky to see as it is a shimmering point of light never far from the Sun. To the untrained eye looking at images of it taken by space probes, it looks a lot like our Moon – an airless, pockmarked, cratered world with mountain ranges and vast plains, baking but, unlike the Moon, underneath a fierce sun that makes the surface hot enough to melt lead. But look closer and there are differences. Mariner 10 was the first space probe to visit this remote planet, which it reached in 1974, and found it to be the second-densest planet in the solar system after the Earth. Its high density suggested a large core of iron and astronomers speculated that when Mercury was young it suffered a big collision that blew away its lighter exterior rocks, diminishing the mantle it was able to form. Despite this, a journey to the heart of this world would be in many ways remarkably similar to the journey through our planet.

Mercury is much smaller than the Earth and could fit well inside the Earth's outer core. It has a solid silicate rock crust about 50 km thick beneath which is a mantle only 200 km in thickness. Then there is a 50-km iron-sulphide barrier followed by a liquid-iron outer core 830 km thick and a solid-iron inner core with a radius of 1,240 km. What this means is that, despite Mercury being much smaller than the Earth – it has a radius only 40 per cent of our own planet – its solid inner core is a little larger than ours. No wonder it is a dense planet! Mercury's core occupies 85 per cent of the planet's radius, making it the largest core relative to the size of the planet in our solar system. It also has a magnetic field that is believed to be generated by motions in its metallic core. Due to its small size the conditions we would experience on a journey to the centre of Mercury are those we have already encountered in our journey to Earth's core. The pressure at its heart is only 11 per cent of that at the centre of the Earth, conditions found not very deep inside the Earth.

Cloud-covered Venus, the next planet out from the Sun, is sometimes considered the Earth's twin, being similar in size, but, as someone once wrote, the doorways to heaven and hell are adjacent and identical. If the Earth is heaven then Venus, with its extremely thick and hot atmosphere that at certain altitudes rains droplets of dilute sulphuric acid, is hell. The similarity in size and density between Venus and Earth suggests they share a similar internal structure and thus a similar journey: a core, mantle and crust, though the pressure at the centre of Venus is 81 per cent that of the centre of the Earth. The main difference between the two planets is the lack of evidence for plate tectonics, possibly because its crust is too strong to subduct without water to make it less viscous. This results in reduced heat loss from the planet, preventing it from cooling and explaining why it does not have a working dynamo utilising convection to generate a magnetic field. Venus's crust appears to be 50 km in thickness, and composed of silicate rocks. Its mantle is approximately 3,000 km thick, but its composition is unknown. It is also not known if its core has solidified. Future spacecraft may deposit seismometers on the surface to look for 'Venusquakes' that will help probe the planet's interior. Only with such data, possibly decades in the future, could we really write the story of a journey to its heart; though I think the evidence so far suggests that any journey to the centre of Venus would not be as scientifically interesting, or as dramatic, as the one to the centre of the Earth.

But what of a journey to the core of the Moon? It is also believed to be composed of a crust, mantle and core. The crust is composed primarily of oxygen, silicon, magnesium, iron, calcium and aluminium and is estimated to be on average about 50 km thick. The lunar mantle is more iron-rich than that of the Earth. Moonquakes – detected by seismometers left behind by the Apollo moonwalkers – have been found to occur deep within the mantle of the Moon about 1,000 km below the surface. These occur with monthly periodicities and are related to tidal stresses caused by the eccentric orbit of the Moon about the Earth. Several lines of evidence imply that the lunar core is small, with a radius of about 350 km or less – only about 20 per cent the size of the Moon itself. The composition of the lunar core is not well constrained by what we know, but most believe that it is composed of metallic iron with a small amount of sulphur and nickel. An analysis of the Moon's variable rotation suggests that the core is at least partly molten. All this considered, a journey to the centre of the Moon would be a brief one and less eventful than a journey to the Earth's core.

Of all the worlds in our solar system Mars is the most Earth-like. It has surface features we would instantly recognise: mountain ranges, craters and canyons, deserts with sand dunes blown by the wind and morning mist that clears from deep valleys. But there is one aspect that makes a voyage to its centre similar to a fragment of the Earth's equivalent journey. That is its size, being about midway between the Earth and the Moon. In fact the Earth's core is about the same size as Mars. Like the Earth, Mars has undergone differentiation, resulting in a dense, metallic core region overlaid by less dense materials. Because of its size the entire mantle on Mars, or almost all of it, is similar to the upper mantle of the Earth and indications are that it should be convecting due to the heat of the core beneath it. What makes Mars different from the Earth is that it does not have tectonic plates but instead has an insulating lithosphere that tends to keep the heat in. In the past Mars appears to have had a dynamo operating in its liquid iron core that generated a magnetic field. This is suggested by observations made by the Mars Global Surveyor that arrived on the red planet in 1997. It detected regions of Mars that had intense magnetic fields that are thought to be remnants of a global magnetic field. It appears that the once liquid core of Mars solidified many billions of years ago, quenching the dynamo. Hence a journey to the core of Mars would not be as exciting as one to its terrestrial counterpart.

If the journey to the centre of the Earth is by far the most interesting journey to the centre of any planet we have so far considered, it meets stiff competition when it comes to a journey to the heart of Jupiter – the solar system's largest planet with a mass of 318 earths. Just looking at Jupiter shows how different such a journey would be, for it has no surface. What you can see is its atmosphere of multi-coloured bands and zones which are jet streams and weather systems larger than the Earth itself. The clouds are composed of ammonia crystals with the cloud layer being about 50 km thick with a thin, clear region below. Water clouds may be present and titanic bolts of lightning a thousand times greater than those on Earth have been detected. Warmer clouds upwell from below and change colour to orange and brown as convection cells carry them up and then down again.

Thus a journey to Jupiter's heart would start in the clouds and descend as the atmosphere of molecular hydrogen became thicker and thicker until it was liquid with no clear boundary between the gaseous and liquid regions. Then you would reach a global ocean of metallic hydrogen. This ocean would have currents and tides and bulge at the equator due to Jupiter's rapid spin. This swirling liquid metal is responsible for generating Jupiter's incredibly strong magnetic field. Chunks of diamonds may be floating amongst the metallic waves and at even lower depths the extreme pressure and temperature might melt them, making it rain liquid diamond. There may even be similar metallic oceans found inside Saturn, on which diamonds may grow so large that they could be called 'diamondbergs'. Diamonds may be even more numerous in the interiors of Uranus and Neptune.

Back inside Jupiter and sailing through this sunless sea, which can reach temperatures in excess of 10,000 degrees C, you eventually arrive at the dense core that is some twelve to forty-five times the mass of the Earth, or between 5 and 15 per cent of the mass of Jupiter. The temperature here is a staggering 37,000 degrees C, with a pressure over ten times that found at the centre of the Earth. Some think the core might be rocky but no one is really sure. It's believed that the rocky or icy core formed first in the solar nebula and started to gather hydrogen around it. It might not even have survived; over time the core may even have dissolved into the sea that surrounds it.

There are also worlds in our solar system whose internal journeys would provide dramatic interludes in a duller trip than that experienced on Earth. Take Europa – an ice-crusted moon of Jupiter with a radius of about a quarter of the Earth. It is believed that underneath its ice shell is an ocean of liquid water in which life could have evolved. Dramatic certainly, and further inside Europa probably has a metallic iron core.

Standard planetary models indicate that the interior of Saturn, the next planet from the Sun, is similar to that of Jupiter, with a small rocky core of 11,700 degrees C, similar in composition to the Earth but more dense. Some estimates put it at nine to twenty-two times the mass of the Earth, which corresponds to a diameter of about 25,000 km. It would be surrounded by a thicker liquid metallic hydrogen layer than found on Jupiter, followed by a liquid layer of helium-saturated molecular hydrogen that gradually transitions into gas with increasing altitude. With conditions more suitable for producing diamonds the journey to the centre of Saturn might prove a more interesting trip.

Next we come to the 'ice giants'. The first, Uranus, has a mass about fourteen times that of the Earth, making it the least massive of the giant planets. Its diameter is slightly larger than Neptune's at roughly four times Earth's. The standard model of its structure is three layers: a rocky silicate/iron-nickel core in the centre, an icy mantle in the middle and an outer gaseous hydrogen/helium en-velope. The core is relatively small, with a mass of only half an Earth mass and a radius less than 20 per cent of Uranus's. According to research the extreme pressure and temperature within Uranus may break up methane mole-cules, with the carbon atoms released forming diamonds that rain down through the mantle like hailstones. Very high-pressure experiments at the Lawrence Livermore National Laboratory using lasers to compress diamonds suggest that the base of the mantle may comprise an ocean of liquid diamond on which float 'diamondbergs'. On a journey through this world, similar to Neptune, one can imagine pausing to land on a slab of diamond the size of a small island and leaving just before it collided with a swarm of diamonds the size of aircraft carriers.

240

But what of journeys through other worlds beyond our solar system? Out there are planets unlike those found in our region of space. In the just over twenty years since the first one was detected astronomers have discovered almost 2,000 planets circling stars out in space. So many that it is now believed there are planets around most of the stars we see. The universe may be teeming with planets, and of course that has implications for the occurrence of life. Calculations have been made concerning the structure of planets with masses several times that of the Earth which would be larger and hotter and have a greater pressure at their cores than our own planet. There is much debate about such worlds and the features they would have and how they would evolve. Some scientists suggest these 'super-Earths' could have more vigorous tectonic activity than the Earth due to more stress, heat flow and thinner plates. Others think that higher gravity would result in a very strong crust and no plate tectonics. Inside such worlds there may be more phase changes to the rocks under greater pressures and probably more internal divisions. Alternatively, the high pressures coupled with large viscosities and high melting temperatures could prevent their interiors from separating into different layers and so result in undifferentiated coreless mantles. Magnesium oxide, which is rocky on Earth, can be a liquid metal at the pressures and temperatures found in super-Earths and could generate a protective magnetic field for billions of years. What forms of life could develop on these high-gravity worlds? I wonder. It is possible that the journey to the centre of a planet whose mass is many times that of the Earth could be more dramatic than our own planet, but it is also possible such worlds could be blander.

27

Journey's End

The Aztecs foretold a time 'when the Earth had become tired . . . when the seed of the earth has ended'.

Africa is set to rotate clockwise and move to the north-west, converging on the Eurasian plate. The Mediterranean will close, its oceanic crust being subducted, forming new mountain ranges as Europe is pushed northward. The great African Rift Valley will stretch to form a new ocean basin, the Atlantic will widen, the Pacific will shrink. In a hundred million years Africa and Europe will merge and the Antarctic will move northwards and reunite with Australia, presaging the formation of the next supercontinent, Pangaea Ultima, in 250 million years' time. When most of the world's continental crust will be concentrated the inner core will still be cooling, and growing, but it will only be about 500 km larger. Perhaps at that stage it might cause perturbations in the way the geodynamo operates in the outer core. A billion and a half years from now the inner core will be twice as large.

No one knows how long the dynamo will continue or how it might be affected by the growing inner core. New models of the geodynamo and the dynamos seen in other planets suggest that the Earth's dynamo could be long-lived – perhaps seven to eight billion years. It is a puzzle why other planets have failed to sustain their dynamos for very long when the Earth's geomagnetic field intensity today is greater than it has been in the past. It is without doubt that among the terrestrial planets the Earth has the finest dynamo. It's the most energetic and we are the only planet with a global system of plate tectonics. Indeed, over the past five million years the geomagnetic field has reversed polarity at near-record rates that are taken as a sign of vigour.

Our Sun is 4.6 billion years old and it has already burned half of its hydrogen fuel at its core. We have hundreds of millions of years of a steady Sun ahead of us. But someday it will change. The Sun will cease to be our friend and the giver of life. For most of our Sun's lifetime it will reside peacefully on the so-called Main Sequence for eleven of the twelve or so billion years of its life. But that is not to say there will not be changes; there will, and they will affect the Earth and mankind profoundly. For the next several billion years the temperature of the Sun's surface, and consequently its brightness, will increase, being about 10 per cent up in the next 1.1 billion years. Some believe that as the brightness of the Sun increases the concentration of water vapour in the Earth's atmosphere will increase as well, rapidly leading to the possibility of a runaway greenhouse effect on our planet that could turn the Earth into another Venus. Some calculations suggest that, as a result of this, in 900 million years the amount of carbon dioxide in our atmosphere will have fallen to a level where plants will have problems surviving. If we lose the plants we are in deep, possibly terminal, trouble. Within the next billion years enhanced ultraviolet radiation could destroy the stratosphere and evaporate the oceans. The Earth could be an inhospitable, uninhabitable wasteland long before the Sun dies. But its core would still be hot.

It could be worse. In three billion years, long before the end of our Sun, the Andromeda galaxy will collide with our own galaxy. Collision is perhaps too strong a word as our two galaxies are mostly empty space; they would pass through each other with stellar collisions being exceedingly rare. More problematic would be the gravitational interaction between the two star systems as they swing around each other in a wide orbit. When this happens to other galaxies we see vast streamers and trails of stars arching away from them as stars are tossed out into intergalactic space. If our Sun is slung out it could end its days not among its fellow stars of the galaxy in which it was born but far away, almost alone, in the vastness of intergalactic space.

It was thought that when our Sun ran out of nuclear fuel it would expand to several hundred times its original diameter and consume its nearby retinue of planets. The swelling is a reaction of the outer layers to significant changes going on deep in the star. But the helium fuel the star is now consuming to maintain itself will not last long, as it is drawing on rapidly diminishing reserves. There is another expansion of the outer layers and the star grows to be a supergiant, becoming much larger and much more luminous than the original Sun used to be. Our Sun will never burn so brightly as it does towards the end.

How would the Earth resist the brightening Sun? In the past the Earth has responded to changes in the Sun's output by adjusting itself so that over billions of years a roughly constant set of conditions existed on its surface.

Some call this the Gaia hypothesis – that the earth will act as a self-regulating system to maintain the conditions for life to exist. Personally, I think that the Gaia idea reads too much anthropomorphism into a non-linear, self-regulating system that has been seen to be stable between certain limits in the past. But whatever your thoughts about the status of the Gaia hypothesis it will not rescue us in the future. The changes that are coming will be beyond the Earth's ability to cope with. In about 7.5 billion years' time the expanded Sun's luminosity will peak at several thousand times what it is today. Then, when too little envelope mass is available to feed the hydrogen-shell burning zone, the Sun's outer layer is puffed off leaving a white dwarf star to cool, almost for ever. But would the Earth have been fried and then swallowed by the expanding Sun?

Detailed calculations suggest that in the later stages of its life the Sun will lose mass and increase its size to a radius of 168 million km, much larger than the 150 million km distance that the Earth orbits the Sun today. No planet can survive for very long when touched by the expanding envelope of its encroaching star; the drag on its orbital motion would doom Mercury and Venus. Once it was thought that the Earth would be spared being swallowed up by the Sun, but that didn't take into account the enormous tidal interaction between the Sun and the Earth, which would quickly rob our world of orbital energy and pull it into the Sun to face its destruction. Just before it dies the Earth will look like Mercury, a wrecked, baked and blasted, bone-dry, scarred hulk with the exposed floors of former oceans. From the lonely ruin that is the Earth at this time the leering red Sun would cover 70 per cent of the sky. The Earth's tidal death will be swift; perhaps in a few hundred years at most it will fragment and scatter itself across the face of the star that created it all those billions of years ago.

The atmosphere would have gone long ago, and what happens next is rather like the journey to the centre of the Earth we have been undertaking. Next would be the strong crust which has not moved across the planet for billions of years, for the great engine of plate tectonics has ceased. Internal radioactive heating has declined and subduction stopped. The Moho will be unpeeled and the mantle transition zone will splinter, trailing behind the dying Earth in an arc of debris. This is no passive dismantling for, as the pressure of the overlying rocks is released, the mantle will explode, buckle and split until the still white-hot metal heart of the Earth is ripped out. It will prove a little more resistant to the ablation in the Sun's atmosphere than the silicate layers, but the outcome is inevitable. No molten iron will be spilled, for the outer core will have been invaded by the solidifying inner core long ago. The core, exposed for the first time, will glow briefly, and then die.

By that time mankind will either have become extinct or

what we have evolved into will have left for who knows where else in the space and time of the cosmos. Perhaps we will leave probes in the vicinity to record what happens; after all, it was the death of the planet that gave us life. Perhaps we will have long forgotten the Earth, and no one will mourn or even notice its passing. When it happens all that will be left will be atoms; every chemical bond that once made a mineral, or a molecule, or a strand of DNA, will be torn apart.

We have seen it happen before. Astronomers have detected white dwarf stars feasting on the remains of Earth-like planets. Normally, a white dwarf's atmosphere is a mix of only hydrogen and helium, because a white dwarf's intense gravity pulls heavier elements toward its core. Because of this, when astronomers detect other elements in a white dwarf's atmosphere they think they must have come from debris falling on to the star's surface.

After a survey of eighty white dwarf stars using the Hubble Space Telescope astronomers found four white dwarf stars whose atmospheres contain oxygen, magnesium, iron, silicon and a small bit of carbon – exactly what was expected if the stars are absorbing dust from former planets. One astronomer mused that if you could shred the Earth into dust and put it into the white dwarf, it would match the chemical composition seen. One of the stars they had observed, called PG0843+516, is even more heavily iron-enriched than the other dwarfs. It contains a high abundance of nickel and sulphur, suggesting that one of the planets PG0843+516 destroyed had an iron–nickel core just like the Earth.

The Earth began as dust clumped together and it will end fragmented into dust scattered across the face of a white dwarf star that will probably do nothing but cool for ever, undisturbed as it moves, constantly fading, in between the stars.

28

Playthings of the Earth

Our journey through the Earth has also taught us something about the search for life in space. It is an obvious thing to say that we know of only one place in the universe where there is life – here on our own planet. It circles an average star, the type found throughout the length and breadth of the cosmos, the type that provides a steady outpouring of energy for billions of years, the type that must shine its light on an almost countless number of Earth-like worlds. Perhaps life's story is taken up on some or many of them, for the vastness of the universe dictates that even if the chances of life developing are low, even almost vanishingly low, then sheer numbers will render it commonplace. But could we ever find that life? Our Earth journey offers some clues.

Ever since the dawn of radio astronomy in the 1940s and 1950s scientists have realised that for the first time in history they possess a device that has a realistic chance of sending and receiving a message over interstellar distances. Radio waves are scarce in the cosmos and travel great distances on relatively little energy, so they are a good choice to send a message. So we can listen, and we have been doing so for over fifty years and so far heard nothing. We might detect an extra-terrestrial message today, or we might have to wait a thousand years, or perhaps we shall never find one. We have a guide to our chances in the form of what is known as the Drake equation, named after its originator, Dr Frank Drake, an American radio astronomer who carried out the first radio-wave searches for a message from extra-terrestrials. It estimates the number of civilisations in our galaxy with which radio communication might be possible, relating that number to the average rate of star formation in our galaxy, the fraction of those stars that have planets, the average number of planets that could potentially support life, the fraction that do support life, those that go on to develop some form of intelligence and those that make the effort to communicate. Take all those numbers and multiply them by the length of time a civilisation will release detectable signals into space and you have the number of civilisations in our galaxy we could theoretically converse with. Because so many factors in the equation are poorly known, that number could range from 1 to 280 million.

Planets are commonplace in the universe, that is one of the major discoveries in astronomy over the past twenty years or so, but what of the possibility of those planets developing and sustaining life? On Earth life seems to have started just as soon as the conditions were suitable, so one could think that life would arise often on other planets. But there must be more to it than that. Our journey through the Earth has shown us just how life on Earth is tied intimately to our planet. The process of subduction, seemingly just a geological process, may be essential for biology to exist on Earth by stimulating the mantle to produce the material and environment necessary for life. Consider the so-called Great Oxygenation Event of 2.3 billion years ago. This was a great increase in the amount of oxygen in the Earth's atmosphere put there by single-celled organisms generating it through photosynthesis. The event may have been related to tectonic changes, as there is some evidence that links it to the formation of continents and that the event occurred suddenly after a gradual but fundamental shift in the way the Earth's crust was accumulating. The Great Oxygenation Event also stimulated a dramatic growth in the number of minerals on Earth, probably more than doubling them to the 4,500 or so found there today. It seems that the dramatic increase in oxygen changed crust and mantle processes. The emergence of the continents would also have increased erosion which, as well as breaking down rocks, removes carbon dioxide from the atmosphere. Carbon dioxide dissolves in rainwater to form carbonic acid that attacks rocks. It is then carried to the sea by streams and rivers with other dissolved minerals distributing important biological nutrients such as iron and phosphorus. Biology, rocks and the geological workings of the Earth are interconnected. This means that the Drake equation needs to take into account the relationship between life and planets, because life just doesn't live on a planet, it *is* part of a planet. Shelley put it well when he said that nothing in the world is single.

We are connected to the outer core through its protective magnetic field, without which our atmosphere would have been stripped away by the solar wind a long time ago, making the surface of our planet at least uninhabitable. One day the inner core will grow to the point where the nature of the geodynamo generation starts to change. If the dynamo falters the Earth's magnetic field will collapse over about 10,000 years, but by the time this happens the Earth will have already destroyed humanity or evicted it.

Jules Verne summed it up very well in *Journey to the Centre of the Earth* when he wrote: 'We were to become mere playthings of the earth.' We playthings live above an inferno, but it is not Dante's inferno. In his famous poem, after visiting hell and seeing the devil encased in ice Dante continues to the very centre of the Earth and climbs upwards to emerge 'beneath a sky studded with stars'. If we are to know ourselves we must appreciate the stars above and also the deep places below where the Sun is silent.

Early in the afternoon of Friday, 24 March 1905 Jules Verne knew that day would be his last. He had been suffering from diabetes and had a stroke recently which had left him half blind. In his last hours he was conscious for a while. His family gathered around him, as did his publisher, though accounts say that Verne did not recognise him.

Jules Verne went to his death harbouring resentment towards the French establishment. He felt ignored, dismissed as a 'popular writer'. Every time he had been nominated for membership of the French Academy he had been rejected. He had never earned as much money from his books as he thought he should have, even though he sold prolifically and influenced generations to take an interest in science. *The Times* obituary said his style of writing was a 'vein [that] was seen to be more or less worked out'.

Scientifically the story has not aged well, although it remains captivating if a little simple for modern tastes. Verne's 'follow-up' to *Journey to the Centre of the Earth* was *The Child of the Cavern*, published in 1877, and it is the last of his novels in which something new could be discovered. The rest of his works, even the very successful *Around the World in Eighty Days*, are repetitions of earlier ideas. Science was moving on and restricting what could be written about the Earth. Sometimes Verne turned his hand towards other subjects. *The Chancellor*, written in 1870, was influenced by the notorious events on the raft of the *Méduse*, when over three-quarters of the people who abandoned ship in shallow waters off the west coast of Africa drowned. It contains no science and little optimism.

Reading *Journey to the Centre of the Earth* it is obvious that it is unlike his other books. It is a journey into the distant past, different from the forward-looking *From the Earth to the Moon*. There are no surviving proofs of *Journey to the Centre of the Earth* and the only reference to it is in a letter of 12 April 1864, the year of its publication. For many years after his death it was believed that there was no extant manuscript of it, but one was discovered in 1994. We do know that the book underwent many changes after it was first published, probably as a result of recent scientific discoveries.

Only 60 per cent of the book actually takes place beneath the surface. Also, the 'to' of the title is misleading, for it is not certain if one can reach the centre. Arne Saknus-semm claimed to have been there but, as Axel points out, reliable ways of measuring depth had not been invented at the time. But key to the book is not the invented caverns or creatures from the Silurian epoch but that Verne pits Axel against Lidenbrock. 'Science, my lad, is made up of mistakes, but they are mistakes which it is useful to make, because they lead little by little to the truth,' he says. Here we have modern science arguing against old-fashioned natural philosophy. I said that *Journey to the Centre of the Earth* was a simple book, and so it is in terms of narrative. But it is not simple in other ways. It is literally turning the world inside out for its readers, and Verne lets us know that any overarching conception of the Earth can be destroyed by a single new find. We see a harmony in this wilderness below us, and it is facts that count, not theories. Things are not always what they seem, like the Earth itself.

Index